T0201978

Franco Strocchi
Scuola Normale Superiore
Classe di Scienze
Piazza dei Cavalieri, 7
I-56100 PISA
Italy

Symmetry Breaking in Classical Systems and Nonlinear Functional Analysis

Franco Strocchi

Symmetry Breaking in Classical Systems
and Nonlinear Functional Analysis

LECTURE NOTES

SCUOLA NORMALE SUPERIORE
1999

Contents

Introduction

These notes essentially reproduce lectures given at the International School for Advanced Studies (Trieste) and at Scuola Normale Superiore (Pisa) on various occasions. The scope of the short series of lectures, typically a fraction of a one-semester course, was to explain on general grounds, also to mathematicians, the phenomenon of Spontaneous Symmetry Breaking (SSB), a mechanism which seems at the basis of most of the recent developments in theoretical physics (from Statistical Mechanics to Many-Body theory and to Elementary Particle theory).

Besides its extraordinary success, the idea of SSB deserves being discussed also because of its innovative philosophical content and in our opinion it should be part of the background knowledge for mathematical and theoretical physics students, especially those who are interested in questions of principle and in general mathematical structures.

By the general wisdom of Classical Mechanics, codified in the classical Noether theorem, one learns that the symmetries of the Hamiltonian or of the Lagrangean are automatically symmetries of the physical system described by it, which does not mean that the (equilibrium) solutions are symmetric, but rather that the symmetry transformation commutes with the time evolution and hence is a symmetry of the physical behaviour of the system . This belief therefore precludes the possibility of describing systems with different dynamical properties in terms of the same Hamiltonian. The realization that this obstruction does not a priori exists, that one may unify the description of apparently different systems in terms of a single Hamiltonian and to account for the different behaviours by the mechanism of SSB, is a real revolution in the way of thinking in terms of symmetries and corresponding properties of physical systems. It is in fact non-trivial to understand how the conclusions of the Noether theorem can be evaded and how a symmetry of the dynamics cannot be realized as a mapping of the physical configurations of the system, which commutes with the time evolution.

The standard folklore explanations of SSB, which one may often find in the literature, is partly misleading, because it does not emphasize the crucial ingredient underlying the phenomenon, namely the need of infinite degrees of freedom. Despite the many popular accounts, the phenomenon of SSB is deep and subtle and it is not without reasons that it has been fully understood only in recent times. The standard cheap explanation identifies the phenomenon with the existence of a degenerate ground (or equilibrium) state, unstable under the symmetry operation, a feature often present even in simple mechanical models (like e.g. a particle on a plane, each point of which defines a ground state unstable under translations), but which is usually not accompanied by a non-symmetric behaviour.

As it will be discussed in these lectures, the phenomenon is rather related to the fact that, for non-linear (dynamical) systems with infinite degrees of freedom, the solutions of the dynamical problem generically fall into classes or "islands", stable under time evolution and with the property that they cannot be related by physically realizable operations. This means that starting from the configurations of a given island one cannot reach the configurations of a different island by physically realizable modifications. The different islands can then be interpreted as the realizations of different physical systems or *different phases* of a system or as *disjoint physical worlds*.

The spontaneous breaking of a symmetry (of the dynamics) in a given island (or phase or physical world) can then be explained as the result of the instability of the given island under the symmetry operation. In fact, in this case one cannot realize the symmetry within the given island, namely one cannot associate to each configuration the one obtained by the symmetry operation.

The existence of such structures is not obvious and in general it involves a mathematical control of the non-linear time evolution of systems with infinite degrees of freedom and the mathematical formalization of the concept of physical disjointess of different islands. For quantum systems, where the mathematical basis of SSB has mostly been discussed, the physical disjointness has been ascribed to the existence of inequivalent representations of the algebra of local observables.

The scope of these lectures is to discuss the general mechanism of SSB within the framework of classical dynamical systems, so that no specific knowledge of quantum mechanics of infinite systems is needed and the message may be suitable also for mathematical students. More specifically the discussion will be based on the mathematical control of the non-linear evolution of classical fields, with *locally* square integrable initial data, with possibly non vanishing limits at infinity.

The mathematical formalization of physical disjointness relies on the constraint of essential localization in space of any physically realizable operation (so that configurations with different limits at infinity belong to disjoint islands). One can in fact show that an island can be characterized by some reference bounded (locally "regular") configuration, having the meaning of the "ground state", and its H^1 perturbations. Each island is therefore isomorphic to a Hilbert space (*Hilbert space sector*).

The stability under time evolution is guaranteed by the condition that the reference configuration satisfies a generalized stationarity condition, i.e. it solves some elliptic problem. Such a condition is in particular satisfied by the time independent solutions and a fortiori by the minima $\overline{\varphi}$ of the potential and the corresponding Hilbert space sectors $\mathcal{H}_{\overline{\varphi}}$ are of the form $\{\overline{\varphi}+\chi, \ \chi \in H^1\}$. The existence of minima of the potential unstable under the symmetry gives therefore rise to islands (or phases or disjoint physical worlds) in which the *symmetry* cannot be realized or, as one says, it is *spontaneously broken.*

This phenomenon is deeply rooted in the non-linearity of the problem and the fact that infinite degrees of freedom are involved. A simple prototype is given by the non-linear wave equation for a Klein-Gordon field $\varphi : \mathbf{R}^s \to \mathbf{R}^n$, with "potential" $U(\varphi) = \lambda(\varphi^2 - a^2)^2$. The model displays some analogy with the mechanical model of a particle in $\mathbf{R}^n$ subject to the potential $U(q) = \lambda(q^2 - a^2)^2$, which can be regarded as the higher dimensional version of the double well potential in one dimension. But the differences are substantial: in the infinite dimensional case of the Klein-Gordon field, each point q has actually become infinite dimensional and in fact each absolute minimum $\overline{\varphi}$, with $|\overline{\varphi}| = a$ describes the infinite set of configurations which have this point as asymptotic limit, equivalently the Hilbert space of configurations which are H^1 modifications of $\overline{\varphi}$. Whereas in the finite dimensional case there is no physical obstruction or "barrier", which prevents from moving from one minimum to the other, in the infinite dimensional case there is no physically realizable operation, which leads from the Hilbert space sector defined by one minimum to that defined by another minimum, because this would require to change the asymptotic limit of the configurations and this is not possible by means of essentially localized operations, the only ones which are physically realizable. Pictorially, one could say that one cannot change the boundary conditions of the "universe" or of the (infinite volume) thermodynamical phase in which one is living.

The realization of the above structures allows to evade part of the conclusions of the classical Noether theorem and to obtain an improved version which also accounts for SSB. In fact, one may prove that the *local conservation law*, $\partial^\mu j_\mu(x) = 0$, associated to a given symmetry of the Hamiltonian or of the Lagrangean gives rise to a *global conservation law* or to a conserved "charge, in a given island, only if the symmetry leaves the island stable. Thus, the *improved version of Noether theorem* still yields the local conservation laws corresponding to the generators of the symmetry group G of the dynamics, but in a given phase or physical world one has the global conservation law only for the generators of the stability group of the given island.

Clearly, if G is the (concrete) group of transformations which commutes with the time evolution, the whole set of solutions of the non-linear dynamical problem can be classified in terms of irreducible representations (or multiplets) of G, but if G is spontaneously broken in a given island defined by the Hilbert space sector $\mathcal{H}_{\overline{\varphi}}$, the latter cannot be the carrier of a representation of the symmetry group G and in particular the elements of $\mathcal{H}_{\overline{\varphi}}$ cannot be classified in terms of multiplets of G. One could think of grouping together solutions corresponding to initial data of the form $\overline{\varphi} + g\chi$, $g \in G$, which might look as candidates for multiplets of G. As a matter of fact, such sets of initial data correspond to representations of a group of transformations which is isomorphic to G, but which does *not* commute with the dynamics and therefore the above from of the initial data does not extend to arbitrary times; thus the above identification of multiplets at the initial time is not stable under time evolution. In fact, the group of transformations which commute with the time evolution corresponds to $\overline{\varphi} + \chi \to g\overline{\varphi} + g\chi$, $g \in G$, which however does not leave $\mathcal{H}_{\overline{\varphi}}$ stable.

Within this approach, it is possible to prove a classical counterpart of the so-called Goldstone theorem, according to which there are massless modes (i.e. solutions of the free wave equation) associated to each broken generator. The theorem proved here provides a mathematically acceptable substitute of the heuristic arguments and corrects the conclusions based on the quadratic approximation of the potential around an absolute minimum.

Explicit example which illustrate how these ideas work in concrete models are discussed in Sect.8.

The material presented in these lectures, largely relies on joint papers with Cesare Parenti and Giorgio Velo, to whom I am greatly indebted, (see the references at the relevant points). An attempt is made to reduce the mathematical details to the minimum required to make the arguments self contained and convincing also for a mathematically minded reader. The required background technical knowledge is kept to a rather low level, so that the lectures might be accessible also to undergraduate students with a basic knowledge of Hilbert space structures.

Symmetry Breaking in Classical Systems

1 Symmetries of a physical system

The realization of symmetries in physical systems has proven to be of help in the description of physical phenomena; it makes possible to relate the behaviour of similar systems and therefore it leads to a great simplification of the mathematical description of Nature.

The simplest concept of symmetry occurs when one realizes that the shape of an object or of a physical system is invariant or symmetric under geometric transformations, like rotations, reflections etc.. More generally, a system is symmetric, under a transformation of the coordinates or of the parameters which identify its configurations, if correspondingly the dynamical behaviour remains unchanged.

To formalize the concept of symmetry we first recall that the *description of a classical physical system* consists in

i) the identification of all its possible configurations $\{S_\gamma\}$, with γ running over an index set of coordinates or of parameters which identify the configuration S_γ

ii) the determination of their time evolution

$$\alpha^t : S_\gamma \rightarrow \alpha^t S_\gamma \equiv S_{\gamma(t)}. \tag{1.1}$$

A *symmetry* g of a physical system is a transformation of the coordinates (or of the parameters) γ, $g : \gamma \rightarrow g\gamma$, which

1) induces an invertible mapping of configurations

$$g : S_\gamma \rightarrow g S_\gamma \equiv S_{g\gamma} \tag{1.2}$$

2) does not change the dynamical behaviour, namely [1]

$$\alpha^t g S_\gamma \equiv \alpha^t S_{g\gamma} \equiv S_{(g\gamma)(t)} = S_{g\gamma(t)} \equiv g \alpha^t S_\gamma. \tag{1.3}$$

For classical canonical systems, condition (1.3) is equivalent to the invariance of the Hamiltonian under the symmetry g (*symmetric Hamiltonian*).

The realization of a symmetry which relates (the configurations of) two seemingly different systems clearly leads to a unification of their description.

Example 1.1 Consider a particle moving on a line, subject to a double well potential, i.e. described by the following Hamiltonian

$$H = \tfrac{1}{2}p^2 + \tfrac{1}{4}\lambda(q^2 - a^2)^2, \tag{1.4}$$

with q, p the canonical coordinates which label the configurations of the particle. Clearly, the reflection $g : q \rightarrow -q$, $p \rightarrow -p$ leaves the Hamiltonian invariant and is a symmetry of the system; obviously, it maps solutions (of the Hamilton equations) into solutions.

---

[1]To simplify the discussion, here we do not consider the more general case in which also the dynamics transforms covariantly under g (like e.g. in the case of Lorentz transformations). For a general discussion of symmetries and of their relevance in physics see R.M.F. Houtappel, H. Van Dam and E.P. Wigner, Rev. Mod. Phys. **37**, 595 (1965).

Now, consider the two classes of solutions corresponding to initial conditions in the neighborhoods of the two absolute minima $q_0 = \pm a$, with $p_0 < \sqrt{\lambda} a^2/2$, respectively and suppose that, by some (artificial) ansatz, in the preparation of the initial configurations one cannot dispose of energies greater than $\lambda a^4/4$. This means that the two classes of solutions correspond to two effectively different "systems", since one cannot go from one to the other by physically realizable operations. The realization that g relates the configurations of the two systems leads to a unified description of them.

For a particle moving on a plane the analog of the double well potential defines a Hamiltonian which is invariant under rotations around the axis (through the origin) orthogonal to the plane and one has a continuous group of symmetries. There is a continuous family of absolute minima lying on the circle $|\vec{q}_0|^2 = a^2$.

2   Spontaneous symmetry breaking

One of the most powerful ideas of modern theoretical physics is the mechanism of spontaneous symmetry breaking. It is at the basis of most of the recent achievements in the description of phase transitions in Statistical Mechanics as well as of collective phenomena in solid state physics. It has also made possible the unification of weak, electromagnetic and strong interactions in elementary particle physics. Philosophically, the idea is very deep and subtle (this is probably why its exploitation is a rather recent achievement) and the popular accounts do not fully do justice to it.

Roughly, spontaneous symmetry breaking is said to occur when a symmetry of the Hamiltonian, which governs the dynamics of a physical system, does not lead to a symmetric description of the physical properties of the system. At first sight, this may look almost paradoxical. From elementary courses on mechanical systems, one learns that the symmetries of a system can be read off by looking at the symmetries of the Hamiltonian, which describes its time evolution; how can then be that a symmetric Hamiltonian gives rise to an asymmetric physical description of a dynamical system?

The cheap standard explanation is that such a phenomenon is due to the existence of a non-symmetric absolute minimum or "ground state", but it appears clear that the mechanism must have a deeper explanation, since the symmetry of the Hamiltonian implies that an asymmetric stable point cannot occur by alone, (the action of the symmetry on it will produce another stable point). Now, the existence of a set of absolute minima related by a symmetry (or "degenerate ground states"), does not imply a non-symmetric physical description. One actually gets a symmetric picture, if the correct correspondence is made between the configurations of the system (and their time evolutions) and such a correspondence is physically implementable, in the sense that if S is a physically realizable configuration, so is each of its transformed ones.

The way out of this argument is to envisage a mechanism by which, if one starts from one non-symmetric absolute minimum (or "ground" state) S_0, there are physical obstructions to reach its transformed ones, $g\,S_0$, by means of physically realizable operations, starting from S_0, so that effectively one gets confined to an asymmetric physical world. The purpose of the following discussion is to make such a rather vague and intuitive picture mathematically precise.

For a classical finite dimensional dynamical system we agree that two configurations are related by *physically realizable operations* if they are connected by a continuous path of configurations, all with finite energy. In this way, one gets a partition of the configurations into classes and given a configuration S, the set of configurations which can be reached from it, by means of physically realizable operations, will be called the *phase* Γ_S, or the "physical world", to which

S belongs. A *symmetry g* will be said to be *physically realized* (or *implementable* or *unbroken*), in the phase Γ, if it leaves Γ stable.

To illustrate the above definitions, we consider a particle moving on a line, subject to a deformed double well potential, still invariant under the reflection $g : q \to -q$, with two absolute minima at $q_0 = \pm a$, but going to infinity as $q \to 0$. Consider now two kind of (one-dimensional) creatures, one living in the valley with bottom $q_0 = a$ and the other in the valley with bottom $q_0 = -a$. The infinite potential barrier prevents from going from one valley to the other (tunnelling is impossible); so that e.g. the people living in the r.h.s. valley do not have access to the l.h.s. valley neither by action on the initial conditions of the particle nor by time evolution. Thus, the operations which are physically realizable (by each of the two kind of people) cannot make transition from one valley to the other and the particle configurations get divided into two phases, labeled by the two minima Γ_a, Γ_{-a}, respectively. The reflection symmetry is not physically realized in each of the two phases. As a matter of fact, even if the particle motion is described by a symmetric Hamiltonian, the particle physical world will look asymmetric to each kind of creatures: the *symmetry* is *spontaneously broken*.

The somewhat artificial example of spontaneous symmetry breaking discussed above is made possible by the infinite potential barrier between the two absolute minima. Clearly, such a mechanism is not available in the case of a continuous symmetry, since then the (absolute) minima are continuously related by the symmetry group and no potential barrier can occur between them (for a concrete example see the two dimensional double well discussed above). In conclusion, for finite dimensional classical dynamical systems, a continuous symmetry of the Hamiltonian is always unbroken.

The often quoted example of a particle in a two dimensional double well potential in thus a rather misleading example of spontaneous breaking of continuous symmetry (it is also an incorrect example in one dimension unless the potential is so deformed to produce an infinite barrier between the two minima). Actually, most of the claimed simple mechanical examples of spontaneous symmetry breaking discussed in the literature are equally misleading.

Clearly, one can take the point of view of calling spontaneous symmetry breaking the occurrence of non-symmetric ground states, but such a definition does not have much to do with the spontaneous symmetry breaking which occurs in elementary particle physics, in Statistical Mechanics etc., because it does not fully account for the phenomenon, by which a symmetry of the dynamics is not shared by the physical description of the system. This is a much deeper phenomenon than the mere existence of non symmetric (minimum) solutions!

For example, for a free particle on a line, each configuration ($q_0 \in \mathbb{R}, p_0 = 0$) is a minimum of the Hamiltonian and it is not stable under translations, but nevertheless one does not speak of symmetry breaking; in fact according to our definition there is only one phase stable under translations.

Similar considerations apply to classical systems which exhibit bifurcation [2] for which, strictly speaking, one does not have spontaneous symmetry breaking as long as the multiple solutions are related by physically realizable operations. As we shall see later, the latter property may fail if one considers the infinite volume (or thermodynamical) limit, in this case giving rise to spontaneous symmetry breaking.

---

[2]D.H. Sattinger, Spontaneous Symmetry Breaking: mathematical methods, applications and problems in the physical sciences, in *Applications of Non-Linear Analysis*, H. Amann et al. eds., Pitman 1981.

3   Symmetries in classical field theory

As the previous discussion indicates, it is impossible to realize the phenomenon of (spontaneous) breaking of a continuous symmetry in classical mechanical systems with a finite number of degrees of freedom. We are thus led to consider infinite dimensional systems, like classical fields.

To simplify the discussion we will focus our attention to the standard case of the non-linear equation

$$\Box \varphi + U'(\varphi) = 0, \tag{3.1}$$

where $\varphi = \varphi(x, t)$, $x \in \mathbb{R}^s$, $t \in \mathbb{R}$, is a field taking values in $\mathbb{R}^n$, (an n-component field), $U(\varphi)$ is the potential, which for the moment will be assumed to be sufficiently regular, and U' denotes its derivative.

A typical prototype is given by

$$U(\varphi) = \tfrac{1}{4}\lambda(\varphi^2 - a^2)^2 \tag{3.2}$$

which is the infinite dimensional version of the double-well potential discussed in Sect.1. The above equation occurs in the description of non-linear waves in many branches of physics like non-linear optics, plasma physics, hydrodynamics, elementary particle physics etc..[3] The above equation will be used to illustrate general structures likely be shared by a large class of non-linear hyperbolic equations.

The solution of the Cauchy problem for the (in general non-linear) equation (3.1), with given initial data

$$\varphi(x, t = 0) = \varphi_0(x), \qquad \partial_t \varphi(x, t = 0) = \psi_0(x), \tag{3.3}$$

provides the classical field $\varphi(x, t)$ described by eq. (3.1).

---

[3]See e.g. G. B. Whitham, *Linear and Non-Linear Waves*, J. Wiley, New York 1974; R. Rajaraman, Phys. Rep. **21 C**, 227 (1975); S. Coleman, *Aspects of Symmetry*, Cambridge Univ. Press 1985.

In analogy with the previous discussion of the finite dimensional systems, a description of the system (3.1) consists in the identification of the class of initial conditions, for which the time evolution is well defined. Deferring the mathematical details, we will now denote by X the functional space within which the Cauchy problem [4] is well posed, i.e. such that for any initial data

$$u_0 = \begin{pmatrix} \varphi_0 \\ \psi_0 \end{pmatrix} \in X \qquad (3.4)$$

there is a unique solution $u(t, x)$ continuous in time (in the topology of X, see below) and belonging to X for any t, briefly $u(t, x) \in C^0(\mathbb{R}, X)$. Thus, X describes the configurations of the system (3.1) and it is stable under time evolution.

In analogy with the finite dimensional case, a *symmetry* of the system (3.1) is an invertible mapping T_g of X onto X, which commutes with the time evolution. To simplify the discussion, we will make the technical assumption that T_g is a continuous mapping (in the X topology) of the form

$$T_g \begin{pmatrix} \varphi(x) \\ \psi(x) \end{pmatrix} = \begin{pmatrix} g(\varphi(x)) \\ J_g(\varphi(x))\psi(x) \end{pmatrix}, \qquad (3.5)$$

---

[4]For an extensive review on the mathematical problems of the non-linear wave equation see M. Reed, *Abstract non-linear wave equation*, Springer-Verlag, Heidelberg 1976. For the solution of the Cauchy problem for initial data not vanishing at infinity, a crucial ingredient for discussing spontaneous symmetry breaking, see C. Parenti, F. Strocchi and G. Velo, Phys. Lett. **59B**, 157 (1975); Ann. Scuola Norm. Sup. (Pisa), III, 443 (1976), hereafter referred as I. A simple account with some addition is given in F. Strocchi, in *Topics in Functional Analysis 1980-81*, Scuola Normale Superiore Pisa, 1982. For a beautiful review of the recent developments see W. Strauss, *Nonlinear Wave Equations*, Am. Math. Soc. 1989.

with g a diffeomorphism of $\mathbb{R}^n$ of class C^2 and J_φ the Jacobian matrix of g. Such symmetries are called *internal symmetries*, since they commute with space and time translations. [5]

Under general regularity assumptions on the potential $U \neq 0$, such that for infinitely differentiable initial data the corresponding solution of eq.(3.1) is of class C^2 in the variables t and x, one gets a sharp characterization of the internal symmetries of the system (3.1).

Theorem 3.1 [6] *Under the above assumption on U, normalized so that $U(0) = 0$, any internal symmetry of the system* (3.1) *is characterized by a g which is an affine transformation*

$$g(z) = Az + a, \tag{3.6}$$

where $a \in \mathbb{R}^n$ and A is an $n \times n$ invertible matrix satisfying

$$A^T A = \lambda \mathbf{1}, \tag{3.7}$$

with A^T the transpose of A and λ a suitable constant. A, a, λ, which depend on g, satisfy the following condition,

$$U(Az + a) = \lambda U(z) + U(a). \tag{3.8}$$

Proof The condition that $T_g \alpha^t u_0 = \alpha^t T_g u_0$ be a solution of eq. (3.1), for any initial data u_0, implies [7]

$$0 = \Box g_k(\varphi) + U'_k(g(\varphi)) =$$

---

[5] For the discussion of more general symmetries see C. Parenti, F. Strocchi and G. Velo, Comm. Math. Phys. **53**, 65 (1977), hereafter referred as II; Phys. Lett. **62B**, 83 (1976).

[6] Ref. II (see above footnote).

[7] We use the convention by which sum over dummy indices is understood; furthermore the relativistic notation is used: $\mu = 0, 1, 2, 3, \partial_0 = \partial/\partial t, \partial_i = \partial/\partial x_i, i = 1, 2, 3, \partial^\mu = g^{\mu\nu}\partial_\nu, g^{00} = 1 = -g^{ii}, g^{\mu\nu} = 0$ if $\mu \neq \nu$.

$$= \frac{\partial^2 g_k}{\partial z_i \partial z_j}(\varphi)\,\partial^\mu \varphi_i \partial_\mu \varphi_j - \frac{\partial g_k}{\partial z_i}(\varphi) U_i'(\varphi) + U_k'(g(\varphi)). \qquad (3.9)$$

Choosing the initial data such that $\varphi_0(x) = const \equiv c, \psi_0(x) = 0$, for x in some region of $\mathbb{R}^s$, the first term of eq.(3.9) vanishes there and one gets

$$-\frac{\partial g_k}{\partial z_i}(c) U_i'(c) + U_k'(g(c)) = 0. \qquad (3.10)$$

Since c is arbitrary, the sum of the last two terms vanishes for any φ. Choosing now $\varphi_0(x) = c, \psi_0(x) = const = b$, for x is some region of $\mathbb{R}^s$, one gets

$$\frac{\partial^2 g_k}{\partial z_i \partial z_j}(c) = 0, \ \forall c \in \mathbb{R}^n, \quad \text{i.e. } g(z) = Az + a.$$

Eq. (3.9) then becomes

$$\frac{\partial}{\partial z_l} U(Az + a) = (A^T A)_{li} \frac{\partial}{\partial z_i} U(z).$$

Since $U \neq 0$ and the above equation must hold for any $z \in \mathbb{R}^n$, it must be $A^T A = \lambda \mathbf{1}$ and therefore $U(Az + a) = \lambda U(z) + const$; the normalization $U(0) = 0$ identifies the latter constant as $U(a)$.

Having characterized the possible symmetries of the classical field (3.1), we may now ask whether we can have spontaneous symmetry breaking. For continuous groups this question seems to be settled by the classical Noether (first) theorem.

Theorem 3.2 *(Noether [8]). Let G be an N parameter Lie group of internal symmetries for the classical system (3.1), then there exist N conserved currents*

$$\partial^\mu J_\mu^\alpha(x, t) = 0, \qquad \alpha = 1, ... N \qquad (3.11)$$

and N conserved quantities

$$Q^\alpha(t) = \int d^s x \, J_0(x, t) = Q^\alpha(0). \qquad (3.12)$$

---

[8]E.Noether, Nachr. d. Kgl. Ges. d. Wiss. Göttingen (1918), p.235.

For the proof we refer to any standard textbook.[9] It is worthwhile to stress that to deduce eq. (3.12) some regularity properties of the solution are needed, even if they are not spelled out in the standard accounts of the theorem.[10]

The above theorem seems to rule out the possibility of symmetry breaking, since it would imply that a continuous symmetry of the Lagrangean or of the Hamiltonian gives rise to a constant of motion which has the meaning of the generator of the symmetry group. In order to answer the deep physical question of whether spontaneous breaking of a continuous symmetry can occur for the classical system (3.1) a more refined knowledge of the structural and mathematical properties of the solutions is needed. In particular, we have to find out whether in the set of solutions one can identify closed "islands" or phases, stable under time evolution (being the infinite dimensional analogue of the valleys of the Example discussed in Sect. 2). As we will see this will lead us to develop a sort of stability theory for infinite dimensional systems.

---

[9]See e.g. H. Goldstein, *Classical Mechanics*, 2nd. ed., Addison-Wesley 1980; E. L. Hill, Rev. Mod. Phys. **23**, 253 (1951).

[10]See e.g. the above quoted book by H. Goldstein.

4 General properties of solutions of classical field equations

The first basic question is to identify the possible configurations of the systems (3.1), namely the set X of initial data for which the time evolution is well defined and which is mapped onto itself by time evolution. In the mathematical language, one has to find the functional space X for which the Cauchy problem is well posed. In order to see this, one has to give conditions on $U'(\varphi)$ and to specify the class of initial data, or equivalently the class of solutions, one is interested in. Here, one faces an apparently technical mathematical problem, which has also deep physical connections.

In the pioneering work by Jörgens [11] and Segal [12] the choice was made of considering those initial data (and consequently those solutions) for which the total "kinetic" energy is finite [13]

$$E_{Kin} \equiv \tfrac{1}{2} \int [(\nabla\varphi)^2 + \varphi^2 + \psi^2] d^s x < \infty. \qquad (4.1)$$

($\psi = \dot\varphi$). From a physical point of view this condition is unjustified and it automatically rules out very interesting cases, like the external field problem, the symmetry breaking solutions, the soliton-like solutions and, in general, all the solutions which do not decrease sufficiently fast at large distances to make the above integral (4.1) convergent. Actually, there is no physical reason why E_{kin} should be finite, since even the splitting of the energy into a kinetic and a

---

[11]K. Jörgens, Mat. Zeit. **77**, 291 (1961).

[12]I. Segal, Ann. Math. **78**, 339 (1963).

[13]Strictly speaking the kinetic energy should not involve the term φ^2. Our abuse of language is based on the fact that the bilinear part of the total energy corresponds to what is usually called the "non-interacting " theory (whose treatment is generally considered as trivial or under control by normal modes analysis). The remaining term in the total energy is usually considered as the true interaction potential.

potential part has ambiguities in it. Therefore, we have to abandon condition (4.1) and we only require that the initial data are *locally smooth* in the sense that

$$\int_V [(\nabla\varphi)^2 + \varphi^2 + \psi^2] d^s x < \infty \qquad (4.2)$$

for any bounded region V (*locally finite kinetic energy*).

As it is usual in the theory of second order differential equations, one may write eq. (3.1) in the first order (or Hamiltonian) formalism, by grouping together the field $\varphi(t)$ and its time derivative $\psi(t) = \dot{\varphi}(t)$ in a two component vector

$$u(t) = \begin{pmatrix} \varphi(t) \\ \psi(t) \end{pmatrix} \equiv \begin{pmatrix} u_1(t) \\ u_2(t) \end{pmatrix}.$$

Equation (3.1) can then be written in the form

$$\frac{du}{dt} = Ku + f(u), \qquad (4.3)$$

with the initial condition

$$u(0) = u_0 = \begin{pmatrix} \varphi_0 \\ \psi_0 \end{pmatrix}, \qquad (4.4)$$

where

$$K = \begin{pmatrix} 0 & 1 \\ \triangle & 0 \end{pmatrix}, \quad f(u) = \begin{pmatrix} 0 \\ -U'(\varphi) \end{pmatrix}. \qquad (4.5)$$

One of the two components of eq. (4.3) is actually the statement that $\psi = \dot{\varphi}$. It is more convenient to rewrite eq. (4.3) as an integral equation which incorporates the initial conditions. To this purpose, we introduce the one parameter continuous group $W(t)$ generated by K and corresponding to the free wave equation (see Appendix A)

$$W(0) = 1, \qquad W(t+s) = W(t)\,W(s) \qquad \forall t, s.$$

Then, the integral form of eq. (4.3) is

$$u(t) = W(t)u_0 + \int_0^t W(t-s)f(u(s))ds. \qquad (4.6)$$

The main advantage of eq. (4.6) is that, in contrast to eq. (4.3), it does not involve derivatives of u and, as we will see, it is easier to give it a precise meaning.

In the first order formalism the condition that the kinetic energy is locally finite reads: $u_{(1)} = \varphi \in H^1_{loc}(\mathbb{R}^s)$, (i.e. $|\nabla\varphi|^2 + |\varphi|^2$ is a locally integrable function); $u_{(2)} = \psi \in L^2_{loc}(\mathbb{R}^s)$. Thus, we assume the following *local regularity condition of the initial data*

$$u \in H^1_{loc}(\mathbb{R}^s) \oplus L^2_{loc}(\mathbb{R}^s) \equiv X_{loc}. \qquad (4.7)$$

The space X_{loc} is equipped with the natural topology generated by the family of *seminorms*

$$\|u\|_V^2 = \int_V ((\nabla\varphi)^2 + \varphi^2)d^sx + \int_V \psi^2 d^sx \qquad (4.8)$$

As in the finite dimensional case, to solve the Cauchy problem we need some kind of Lipschitz condition [14] on the potential; in agreement with the local structure discussed above, it is natural to chose the following

Local Lipschitz condition:
a) $f(u)$ defines a continuous mapping of X_{loc} into X_{loc}
b) for any sphere Ω_R, of radius R, and for any $\rho > 0$, there exists a constant $C(\Omega_R, \rho)$, such that

$$\|f(u_1) - f(u_2)\|_{\Omega_R} \leq C(\Omega_R, \rho)\|u_1 - u_2\|_{\Omega_R}, \qquad (4.9)$$

for all $u_1, u_2 \in X_{loc}$ such that $\|u_i\|_{\Omega_R} \leq \rho, i = 1, 2$ and

$$\sup_{0 \leq t \leq R/2} C(\Omega_{R-t}, \rho) \equiv \bar{C}(\Omega_R, \rho) < \infty.$$

---

[14]See e.g. V. Arnold, *Ordinary Differential Equations*, Springer 1992, Ch. 4; G. Sansone and R. Conti, *Non-linear Differential Equations*, Pergamon Press 1964.

The above local Lipschitz condition is satisfied by a large class of potentials U:

i) in $s = 1$ dimension, if $U(\varphi)$ is an entire function;

ii) for $s = 2$, if

$$U(\varphi) = \sum_{\alpha \in N^n} C_\alpha \varphi^\alpha, \tag{4.10}$$

α being a multi-index, $\varphi^\alpha = \varphi_1^{\alpha_1}...\varphi_n^{\alpha_n}$, with

$$\sum_{\alpha \in N^n} |C_\alpha| |\alpha|^{|\alpha|/2} |\varphi|^{|\alpha|} < \infty,$$

iii) for $s = 3$, if U is a twice differentiable real function such that

$$\sup_{\varphi} (1 + |\varphi|^2)^{-1} |U''(\varphi)| < \infty. \tag{4.11}$$

The proof that the above classes of potentials satisfy the local Lipschitz condition is similar to that for global Lipschitz continuity (see Lemma 5.3) below), except that local Sobolev inequalities are used instead of global ones (for details see Ref. I, quoted in Sect. 3).

Since, for the present purposes, we are not interested in optimal conditions, (for a more general discussion see Ref. I), in the following, for simplicity, we will consider potentials belonging to the above classes, for $s = 1, 2, 3$.

The above Local Lipschitz condition guarantees that

1) eq. (4.6) is well defined for $u \in C^0(\mathbb{R}, X_{loc})$

2) the *solution* of eq.(4.6), if it exists, *is unique*

3) eq. (4.6) has an *hyperbolic character*, i.e. the local norm of $u(t)$ in the sphere Ω_{R-t} of radius $R-t$, $0 < t < R$, depends only on the local norm of $u(0)$ in the sphere Ω_R of radius R (the influence domain)

$$\|u(t)\|_{\Omega_{R-t}} \le A e^{\omega t} \|u(0)\|_{\Omega_R}, \tag{4.12}$$

(ω a suitable constant)

4) *solutions* of eq. (4.6) *exist for sufficiently small times.*

For the proof of 1)-4), see Appendix B.

To continue the solutions from small times to all times, and in this way get a *global in time solution of the Cauchy problem,* one needs a bound which implies that the norm of $u(t)$ stays finite. This is guaranteed if U satisfies the following

Lower Bound condition: There exist suitable non-negative constants α, β such that

$$U(\varphi) \geq -\alpha - \beta|\varphi|^2. \tag{4.13}$$

In conclusion we have

Theorem 4.1 *(Cauchy problem: global existence of solutions)*[15]. *If U is such that the local Lipschitz condition and the lower bound condition are satisfied, then eq. (4.6) has one and only one solution $u(t) \in C^0(\mathbb{R}, X_{loc})$.*

For a brief sketch of the proof see Appendix C.

---

[15]To our knowledge the proof of global existence of solutions of eq. (4.6) for initial data in $H^1_{loc} \oplus L^2_{loc}$ first appeared in Ref. I, although the validity of such a result was conjectured by W. Strauss, Anais Acad. Brasil. Ciencias **42**, 645 (1970), p. 649, Remark: "The support restrictions on $u_0(x), u_1(x), F(x, t, 0)$ could probably be removed by exploiting the hyperbolic character of the differential equation ...".

5   Stable structures, Hilbert sectors, Phases

The mathematical investigation of the existence of solutions for the non-linear eq.(4.6) does not exhaust the problem of the physical interpretation of the corresponding classical field theory. For infinitely extended systems, in general not every solution is physically acceptable and one has to supplement the analysis of the possible solutions by a list of mathematical properties which the solutions must share in order to make the physical interpretation possible.

For quantum field theory the realization of the basic mathematical structure which renders the theory physically sound is due to Wightman [16] and it is nowadays standard to define as quantum field theories those theories defined by the "solutions" of the quantum field equations which satisfy Wightman's axioms. A similar problem arises in Statistical Mechanics and the basic structure has been clarified [17].

It is then natural that a possible classical field theory associated to the eq. (4.6) be defined by a set S of solutions satisfying a few (additional) basic requirements. General considerations suggest the following ones

I (**Local structure**) A possible classical field theory, or a physical world, associated to the eq. (4.6), is defined by a set S of configurations of the classical field which are related by physically realizable operations (see the analogous property discussed in Sect. 2 and the more precise discussion below).

---

[16]R.F. Streater and A.S. Wightman, *PCT, Spin and Statistics and All That*, Benjamin-Cumming Pubbl. C. 1980.

[17]See e.g. D. Ruelle, *Statistical Mechanics*, Benjamin 1969; R. Haag, *Local Quantum Theory*, Springer-Verlag 1992.

II(**Stability**) S is stable under time evolution

III (**Finite energy-momentum**) An energy-momentum density can be defined in S and its infinite volume integral is finite for each element of S.

To be more precise we have to give a mathematical formalization of the above requirements.

I. **Local structure**. The first condition is based on the physical consideration that our measuring apparatuses extend over bounded regions of space and therefore, starting from a given field configuration u, by physically realizable operations we can modify it essentially only locally, i.e. we can reach only those configurations which essentially differ from u only locally (*quasi local modifications*). From a mathematical point of view it is natural to identify the concept of quasi local modification as a $H^1(\mathbb{R}^s) \oplus L^2(\mathbb{R}^s)$ perturbation, i.e. given a solution $u_1(t)$, a solution $u_2(t)$ is a quasi local modification of u_1 if $u_1(t) - u_2(t) \in H^1(\mathbb{R}^s) \oplus L^2(\mathbb{R}^s)$ continuously in t, briefly

$$u_1(t) - u_2(t) \in C^0(\mathbb{R}, H^1(\mathbb{R}^s) \oplus L^2(\mathbb{R}^s)). \qquad (5.1)$$

We are thus led to introduce the following

Definition 5.1 *Let $\mathcal{F}$ denote the family of solutions $u(t)$ of eq. (4.6), characterized in Sect. 4, $u(t) \in C^0(\mathbb{R}, X_{loc})$, a subset S of $\mathcal{F}$ defines a (essentially) local structure if $\forall u_1, u_2 \in S$, eq. (5.1) holds.*

II. **Stability**. Since time evolution is one of the possible realizable "operations", the above definition of local structure is physically meaningful provided it is stable under time evolution, namely if $u(t) \in S$ also $u_\tau(t) \equiv u(t+\tau) \in S$, $\forall \tau \in \mathbb{R}$. A local structure satisfying such stability under time evolution will be called a **sector**. Thus, all the elements u of the sector S identified by the reference element

$\bar{u}$ have the property that $\delta(t) \equiv u(t) - \bar{u}(0) \in H^1(\mathbb{R}^s) \oplus L^2(\mathbb{R}^s)$, $\forall t \in \mathbb{R}$.

In general S does not have a linear structure, nor that of the affine space $\bar{u}(0) + H^1 \oplus L^2$ since it is not guaranteed that for all $\delta_0 \in H^1 \oplus L^2$, the solution $u(t)$ corresponding to the initial data $\bar{u}(0) + \delta_0$ will belong to S. A sector with such a property is isomorphic to a Hilbert space and it is called a **Hilbert space sector** (HSS).

The above definition of sectors is motivated by simple physical considerations, but since it involves the knowledge of time evolution, it is not obvious how to verify it a priori. The obvious questions are:

i) given a non-linear equation (4.6), can one a priori decide whether there are non-trivial sectors associated to it? In particular, without having to solve eq. (4.6), under which conditions (if any) an initial data define a sector and what is its explicit content?

ii) is it possible that a fully non-linear problem (no approximation being involved) exhibits the existence of linear, actually Hilbert space, structures (in the set of solutions)?

We defer the discussion of condition III to the next section. Now, we prefer to discuss the mathematical structures associated to the above definitions and in particular to show that under general conditions they are not void.

It is not difficult to recognize the analogies with the stability theory, which plays a crucial rôle in the theory of non-linear phenomena, in the finite dimensional case. [18]

---

[18] G. Sansone and R. Conti, *Non-Linear Differential Equations*, Pergamon Press 1964, Ch. IX.

As it appears also in other fields, the concept of "locality" plays an important rôle for the infinite dimensional generalization of ideas developed for finite dimensional systems. The emphasis on local structures is actually the key, which makes possible (and physically meaningful) the treatment of the dynamics of infinite degrees of freedom. Guided by these considerations, we are led to consider the following stability problem: if two configurations $u_1(0)$, $u_2(0)$ are "close" at $t = 0$, in the sense that they differ by a quasi local perturbation, namely $u_1(0) - u_2(0) \in H^1(\mathbb{R}^s) \oplus L^2(\mathbb{R}^s)$, under which conditions they remain "close" at any later times (and therefore are elements of a local structure)?

Clearly, every solution $u(t) \in \mathcal{F}$ defines a local structure (at worst that consisting of just one element), but in general it does not define a sector. In the latter case, the time evolution has a somewhat catastrophic character, since it drastically changes the large distance behaviour of the initial data; as we will discuss below this would mean a change from one physical world to another and this makes a reasonable physical interpretation difficult. Clearly, it is important to have general criteria for the existence of non-trivial stable structures without having to know all the solutions of the non-linear equation.

For simplicity, we discuss the case in which the potential $U(\varphi)$ belongs to the following classes: it is an entire function in dimension $s = 1$ and it belongs to the classes (4.10) and (4.11) in dimension $s = 2, 3$, respectively. For a more general discussion see Ref. II.[19] Then we have

---

[19]C. Parenti, F. Strocchi and G. Velo, Phys. Lett. **62B**, 83 (1976); Comm. Math. Phys. **53**, 65 (1977); Lectures at the Int. School of Math. Phys. Erice 1977, in *Invariant wave equations*, G. Velo and A. S. Wightman eds., Springer-Verlag 1978.

Theorem 5.2 *An initial data*

$$u_0 = \begin{pmatrix} \varphi_0 \\ \psi_0 \end{pmatrix} \in H^1_{loc} \oplus L^2_{loc}.$$

with φ_0 bounded, defines a non-trivial sector $\mathcal{H}_{u_0}$ iff

a) $\psi_0 \in L^2(\mathbb{R}^s),$ (5.2)

b) $\Delta\varphi_0 - U'(\varphi_0) \equiv h \in H^{-1}(\mathbb{R}^s),$ (5.3)

(i.e. the Fourier transform $\tilde{h}(k)$ satisfies $\int |\tilde{h}(k)|^2 (1+k^2)^{-1} d^s k < \infty$).

Actually, $\mathcal{H}_{u_0}$ is completely specified as the set of all solutions $v(t)$ with initial data of the form

$$v_0 = \begin{pmatrix} \varphi_0 + \chi \\ \psi_0 + \zeta \end{pmatrix}, \begin{pmatrix} \chi \\ \zeta \end{pmatrix} \in H^1(\mathbb{R}^s) \oplus L^2(\mathbb{R}^s),$$ (5.4)

i.e. $\mathcal{H}_{u_0}$ is the affine space $u_0 + H^1(\mathbb{R}^s) \oplus L^2(\mathbb{R}^s)$ and thus, being isomorphic to $H^1(\mathbb{R}^s) \oplus L^2(\mathbb{R}^s)$, it carries a Hilbert space structure (**Hilbert space sector**).

Proof Let $v(t)$ be a solution $\in \mathcal{F}$ and $u_0 \equiv \begin{pmatrix} \varphi_0 \\ \psi_0 \end{pmatrix}$, then

$$\delta(t) = \begin{pmatrix} \chi(t) \\ \zeta(t) \end{pmatrix} \equiv v(t) - u_0$$ (5.5)

satisfies the following integral equation

$$\delta(t) = W(t)\delta_0 + L(t) + \int_0^t ds \, W(t-s) \, g(\delta(s)),$$ (5.6)

where

$$L(t) = (W(t) - 1)v_0 + \int_0^t ds \, W(t-s) \begin{pmatrix} 0 \\ -U'(\varphi_0) \end{pmatrix}$$ (5.7)

$$= \begin{pmatrix} \frac{1-\cos\sqrt{-\Delta}\,t}{-\Delta} & \frac{\sin\sqrt{-\Delta}\,t}{\sqrt{-\Delta}} \\ \frac{\sin\sqrt{-\Delta}\,t}{\sqrt{-\Delta}} & \cos\sqrt{-\Delta}\,t - 1 \end{pmatrix} \begin{pmatrix} \Delta\varphi_0 - U'(\varphi_0) \\ \psi_0 \end{pmatrix} \equiv \begin{pmatrix} L_1(t) \\ L_2(t) \end{pmatrix},$$

$$g(\delta(s)) \equiv \begin{pmatrix} 0 \\ -G'(\chi(s)) \end{pmatrix}, \tag{5.8}$$

$$G(\chi) \equiv U(\varphi_0 + \chi) - U(\varphi_0) - U'(\varphi_0)\chi. \tag{5.9}$$

The explicit dependence of G on x through φ_0, $G(x, \chi(x))$, will be spelled out in the following only when necessary. Furthermore, for brevity $\nabla_z G(x, z)|_{z=\chi}$ will be denoted by $G'(\chi)$.

The crux of the argument is that for φ_0 bounded, briefly$\in L^\infty(\mathbb{R}^s)$, for the class of potentials under consideration, $G(\chi)$ satisfies

i) $G'(\chi)$ is *globally Lipschitz continuous*, namely for any $\rho > 0$, there exists a constant $C(\rho)$ such that for any $\chi_1, \chi_2 \in H^1(\mathbb{R}^s)$, with $\|\chi_i\|_{H^1} \leq \rho, i = 1, 2$,

$$\|G'(\chi_2) - G'(\chi_1)\|_{L^2} \leq C(\rho)\|\chi_2 - \chi_1\|_{H^1} \tag{5.10}$$

ii) G satisfies a *lower bound condition*, i.e. there exists a non negative constant γ, such that

$$G(x, z) \geq -\gamma|z|^2, \quad \forall z \in \mathbb{R}^n, x \in \mathbb{R}^s \tag{5.11}$$

(The proof of i) and ii) is given in Lemma 5.3 and 5.4, respectively).

Now, if i), ii) hold, since $g(0) = 0$, property i) implies that $g(\chi) \in H^1(\mathbb{R}^s) \oplus L^2(\mathbb{R}^s)$ and therefore, since $W(t)$ maps $H^1(\mathbb{R}^s) \oplus L^2(\mathbb{R}^s)$ into itself continuously in t, (see Appendix A),

$$\delta(t) \in C^0(\mathbb{R}, H^1 \oplus L^2) \quad \text{iff} \quad L(t) \in C^0(\mathbb{R}, H^1 \oplus L^2). \tag{5.12}$$

The latter condition is equivalent to conditions a) and b), (eqs. (5.2) (5.3)), (see Lemma 5.3 below).

The proof that the sector is not empty and actually is a Hilbert space sector, amounts to proving that eq. (5.6) has one and only one solution $\delta(t) \in C^0(\mathbb{R}, H^1 \oplus L^2)$ for any initial data $\delta_0 \in H^1(\mathbb{R}^s) \oplus L^2(\mathbb{R}^s)$.

A simple but important case is when u_0 is a static solution of eq. (4.6), namely

$$\Delta\varphi_0 - U'(\varphi_0) = 0, \qquad \psi_0 = 0. \tag{5.13}$$

In this case $L(t) = 0$ and eq. (5.6) has the same form of eq. (4.6), for which the Cauchy problem in $H^1 \oplus L^2$ has been solved by Segal [20].

In the general case $L(t) \neq 0$ a generalization of Segal theorem (see Appendix D) gives existence and uniqueness in $H^1 \oplus L^2$.

Lemma 5.3 *For any* $\varphi_0 \in L^\infty(\mathbb{R}^s)$, *the function* $G'(\chi)$ *defined through eq. (5.9) is globally Lipschitz continuous (eq. (5.10)).*

Proof From the identity

$$G'(\chi^{(2)}) - G'(\chi^{(1)}) = U'(\varphi_0 + \chi^{(2)}) - U'(\varphi + \chi^{(1)})$$

$$= \int_0^1 d\sigma \frac{d}{d\sigma} U'(\varphi_0 + \chi^{(2)} + \sigma(\chi^{(2)} - \chi^{(1)}))$$

$$= \int_0^1 d\sigma \, U''(\varphi_0 + \chi^{(2)} + \sigma(\chi^{(2)} - \chi^{(1)}))(\chi^{(2)} - \chi^{(1)}),$$

eq.(5.10) will follow if, for any $\rho > 0$, there exists a constant $C(\rho)$ such that

$$\sup_{k=1,\dots n} \left\| \sum_{j=1}^n \frac{\partial^2 U}{\partial z_j \partial z_k}(\varphi_0 + \chi')\chi_j \right\|_{L^2} \leq C(\rho)\|\chi\|_{H^1}, \tag{5.14}$$

for all $\chi', \chi \in H^1$ with $\|\chi'\| \leq \rho, \|\chi\| \leq \rho$.

For the class of potentials under consideration, the proof of eq. (5.14) reduces to estimating terms of the type $(\varphi + \chi^{(1)})^\alpha \chi^{(2)}$ with

---

[20] See footnote 12.

$\chi^{(i)} \in H^1, i = 1, 2, \alpha \in \mathbb{N}^n$ for $s = 1, 2$ and $|\alpha| \leq 2$ for $s = 3$. Now, since $|a + b|^p \leq 2^p(|a|^p + |b|^p), \forall a, b \in \mathbb{R}, p \geq 1$,

$$\|(\varphi_0 + \chi^{(1)})^\alpha \chi^{(2)}\|_{L^2} \leq 2^{|\alpha|}\{\| |\varphi_0|^{|\alpha|} |\chi^{(2)}| \|_{L^2} + \| |\chi^{(1)}|^{|\alpha|} |\chi^{(2)}| \|_{L^2}\} \tag{5.15}$$

and the first term on the r.h.s. is immediately estimated by

$$2^{|\alpha|}\| |\varphi_0|^{|\alpha|} |\chi^{(2)}| \|_{L^2} \leq A^{|\alpha|}(\|\varphi_0\|_{L^\infty})^{|\alpha|} \|\chi^{(2)}\|_{H^1}. \tag{5.16}$$

The second term can be estimated by using the usual Hölder and the Sobolev inequalities

$$2^{|\alpha|}\| |\chi^{(1)}|^{|\alpha|}|\chi^{(2)}| \|_{L^2} \leq 2^{|\alpha|}\| |\chi^{(1)}| \|^{|\alpha|}_{L^{2(|\alpha|+1)}}\| |\chi^{(2)}| \|_{L^{2(|\alpha|+1)}}$$
$$\leq B^{|\alpha|} C_s(2|\alpha| + 2)^{|\alpha|+1}\|\chi^{(1)}\|^{|\alpha|}_{H^1}\|\chi^{(2)}\|_{H^1}. \tag{5.17}$$

(The derivation of such inequalities is standard.[21])
Thus for $s = 3$ the proof is completed. For $s = 1, 2$ the convergence of the sum over α is guaranteed by the properties which characterize the class of potentials under considerations.

Lemma 5.4 *For $\varphi_0 \in L^\infty(\mathbb{R}^s)$, the lower bound condition for the potential, eq.(4.13), implies that eq. (5.12) holds.*

---

[21]See e.g. L. R. Volevic and B. P. Paneyakh, Russian Math. Surveys **20**, 1 (1965). We list them for the convenience of the reader

$s = 1, \quad \|f; L^p(\mathbb{R}^1)\| \leq C_1(p) \|f; H^1(\mathbb{R}^1)\|, \quad 2 \leq p \leq \infty, C_1(p) = 0(1),$
$s = 2, \quad \|f; L^p(\mathbb{R}^2)\| \leq C_2(p) \|f; H^1(\mathbb{R}^2)\|, \quad 2 \leq p < \infty, C_2(p) = 0(p^{\frac{1}{2}}),$
$s = 3, \quad \|f; L^p(\mathbb{R}^3)\| \leq C_3(p) \|f; H^1(\mathbb{R}^3)\|, \quad 2 \leq p \leq 6, \quad C_3(p) = 0(1).$

The same kind of estimates hold locally. In particular, for any cube $K \subset \mathbb{R}^s$ of size R, they take the form

$$\|f; L^p(K)\| \leq C_{s,R}(p) \|f; H^1(K)\|,$$

with $p \in [2, +\infty]$ for $s = 1$, $p \in [2, +\infty[$ for $s = 2$ and $p \in [2, 6]$ for $s = 3$. The constants $C_{s,R}(p)$ depend only on the size R and exhibit the same dependence on p as in the global case.

Proof Consider the identity

$$G(y) = \int_0^1 d\sigma(1-\sigma)\frac{d^2}{d\sigma^2}U(\varphi_0 + \sigma y) = \int_0^1 d\sigma(1-\sigma)y^2 U''(\varphi_0 + \sigma y).$$
(5.18)

Since U is of class C^2, and φ_0 is bounded, $U''(\varphi_0 + \sigma y)$ is bounded below for $|y| \leq 1$, $0 \leq \sigma \leq 1$; hence from eq. (5.14) we get a lower bound for G of the form of eq. (5.11). Now, for $|y| \geq 1$, the lower bound condition eq. (4.13), gives

$$G(y) \geq - \{\alpha + \beta + \beta \sup_{x \in \mathbb{R}^s}[|\varphi_0(x)|^2 + 2|\varphi_0(x)| + U'(\varphi_0(x))]$$

$$+ \max(0, \sup_{x \in \mathbb{R}^s} U(\varphi_0(x)))\}|y|^2$$

Lemma 5.5 $L(t) \in C^0(\mathbb{R}, H^1 \oplus L^2)$ *iff a) and b) hold.*

Proof Sufficiency is easily seen in Fourier transform, by noticing that

$\cos|k|t - 1$, $(1 + |k|)\sin|k|t/|k|$ and $(1 + |k|)^2|k|^{-2}(\cos|k|t - 1)$ are multipliers of L^2 continuous in t.

For the necessity, we note that $L_2(t) \in C^0(\mathbb{R}, L^2)$ implies that also $\int_0^t d\tau L_2(\tau) \in C^0(\mathbb{R}, L^2)$ and therefore

$$L_1(t) + \int_0^t d\tau L_2(\tau) = -t\tilde{\psi} \in L^2, \quad \text{i.e. } \tilde{\psi} \in L^2.$$

Hence, $|k|^{-1}\sin|k|t \, \tilde{\psi} \in C^0(\mathbb{R}, H^1)$ and the condition on $L_1(t)$ yields

$$f(k,t) = |k|^{-2}(1 - \cos|k|t)\tilde{h}(k) \in C^0(\mathbb{R}, H^1),$$
(5.19)

which in turn implies

$$(|k|^{-2}\sin|k| - |k|^{-1})\tilde{h} = \int_0^t d\tau f(k,\tau) \in C^0(\mathbb{R}, H^1).$$
(5.20)

Finally, the two estimates

$$\tfrac{1}{4}t^2|\tilde{h}(k)| \leq |k|^{-2}(\cos|k|t - 1)|\tilde{h}|,$$

for $|k| \leq 2, t$ sufficiently small, and

$$\tfrac{1}{2}|k|^{-1}|\tilde{h}(k)| \leq (|k|^{-2}\sin|k| - |k|^{-1})|\tilde{h}|,$$

for $|k| \geq 2$, imply $|\tilde{h}|(1 + |k|^2)^{-1/2} \in L^2$, by eqs. (5.19) (5.20).

Remark The conclusions of the above theorem hold in the more general case in which the condition $\varphi_0 \in L^\infty(\mathbb{R}^s)$ is replaced by that of φ_0 being such that i) and ii) (eqs. (5.10) (5.11)) hold; in this case φ_0 is said to be a *regular point* (or admissible) with respect to U. For the discussion of this more general case see Ref. II.

A distinguished case for the application of the theorem is given by the so-called *static solutions*, eq. (5.13), since they define sectors containing a time invariant element. Even more relevant is the case of sectors defined by constant solutions corresponding to absolute minima of the potential; they are the analogue of the *vacuum sectors* of quantum field theory and we will call them *phases*. The constant solutions corresponding to relative minima of U are the analog of the *false vacua* [22] and are classically stable (no tunnelling).

In general, a sector $\mathcal{H}_{u_0}$ identified by a u_0 satisfying the assumptions of Theorem 5.2, does not contain static solutions; a necessary and sufficient condition is that the elliptic equation

$$\Delta\chi - G'_{\varphi_0}(x, \chi) = h(\varphi_0)$$

with $h(\varphi_0) \equiv \Delta\varphi_0 - U'(\varphi_0) \in H^{-1}(\mathbb{R}^s)$, has solutions $\chi \in H^1(\mathbb{R}^s)$.

The occurrence of linear stable structures in a fully non-linear problem is a rather remarkable feature, since each Hilbert space sector is made of solutions of the full non-linear equation, without any approximation or linearization being involved. The *generalized extremal solutions*, namely those corresponding to initial data satisfying the assumption of Theorem 5.2, play a hierarchical rôle since they

---

[22]S. Coleman, Phys. Rev. D **15**, 2929 (1977).

keep their $H^1 \oplus L^2$ perturbations steadily trapped around them. The non-linearity of the problem is (to a large extent) taken care of by the generalized extremal solution φ_0, which defines $\mathcal{H}_{\varphi_0}$, any other element belongs to the affine space generated by φ_0 through $H^1 \oplus L^2$. In general, the time evolution is not described by a linear operator on $\mathcal{H}_{\varphi_0}$.

The occurrence of Hilbert space sectors in the set solutions of non-linear field equations allows to establish strong connections with quantum mechanical structures and to recover the analog of quantum mechanical phenomena like linear representations of groups, spontaneous symmetry breaking, pure phases, superselection rules, etc., at the level of classical equations.[23]

It is worthwhile to remark that the emergence of the above stable structures in the set of solutions of the non-linear eq. (4.6) has been made possible by the framework adopted in Sect. 4, in which the Cauchy data were not restricted to be in $H^1 \oplus L^2$. In that case one would have only gotten the sector corresponding to the *trivial vacuum* $\varphi_0 = 0$, $\psi_0 = 0$.[24]

The physical relevance of such structures should be pretty evident as a consequence of the discussion in Sect. 5 : a phase can in fact be considered as the definition of the "world" of configurations which are physically accessible, starting from a given ground state configuration. By definition of local structure, configurations belonging to the same phase or "world" differ by quasi local perturbations, i.e. they have the same large distance behaviour (for a more detailed discussion see Appendix E); then, since we cannot modify the large

---

[23]F. Strocchi, Lectures at the Workshop on *Recent Advances in the Theory of Evolution Equations*, ICTP Trieste 1979, published in *Topics in Functional Analysis 1980-81*, Scuola Normale Superiore, Pisa 1982; contribution to the Workshop on *Hyperbolic Equations* (1987), published in Rend. Sem. Mat. Univ. Pol. Torino, Fascicolo speciale 1988, pp. 231-250.

[24]See footnotes 11, 12.

distance behaviour of our (reference or) ground state, nor we can change the boundary conditions of our physical world or "universe" by means of physically realizable operations, different phases define *disjoint physical worlds*. The occurrence of disjoint physical worlds or phases is a typical feature of infinitely extended systems, like e.g. those defined by the thermodynamical limit in Statistical Mechanics, for which one cannot go from one phase to another by essentially local operations.[25]

---

[25]The physical relevance of locality has been emphasized by R. Haag and D. Kastler, J. Math. Phys. **5**, 848 (1964) see also R. Haag, loc. cit. (see footnote 17).

6 Sectors with energy-momentum density

We will now discuss the requirement III (Finite energy-momentum) mentioned in the previous Section. Clearly, the possibility of using solutions of non-linear field equations, for the description of physical systems, requires that such solutions have finite energy-momentum and the localization properties of the physical measurements requires the existence of an energy momentum density.

The conventional expression of the energy density for the theory described by eq. (4.6) is

$$\mathcal{E}(\varphi, \psi) = \tfrac{1}{2}[(\nabla \varphi)^2 + \psi^2] + U(\varphi). \qquad (6.1)$$

However, if one adds any function of x, the (Hamilton) equations of motion will remain unchanged and the new expression of the total energy is still formally conserved. This ambiguity is related to the fact that only energy differences have a physical meaning, so that the concept of finite energy solutions must necessarily make reference to some chosen solution. Such a fixing of the energy scale will in general depend on the sector, since $\mathcal{E}(\varphi, \psi)$ is locally but in general not globally integrable, and it corresponds to the so-called *infinite volume renormalization*.

Thus, given a Hilbert space sector (HSS), which from now on we will consider defined by a $\varphi_0 \in L^\infty(\mathbb{R}^s)$ and denoted by $\mathcal{H}_{\varphi_0}$, one is led to define a *renormalized energy density* (without loss of generality we can take $\psi_0 = 0$)

$$\mathcal{E}_{ren}(\varphi, \psi) \equiv \mathcal{E}(\varphi, \psi) - \mathcal{E}(\varphi_0, 0)$$

$$= \tfrac{1}{2}[(\nabla \chi)^2 + \psi^2] + \nabla \chi \nabla \varphi_0 + G(\chi) + U'(\varphi_0)\chi, \quad (6.2)$$

where $\chi = \varphi - \varphi_0$ and $G(\chi)$ is defined by eq. (5.9).

The background subtraction is however not enough for assuring that the renormalized density is globally integrable. The most which can be said, without additional assumptions, is that $\mathcal{E}_{ren}$ is integrable if χ is of compact support and that it identifies an energy functional defined on the whole HSS by a suitable extension [26]. However, in general the so extended functional will not be the integral over a density and therefore the concept of local energy is problematic. Such a difficulty does not arise if the HSS is defined by a $\varphi_0 \in L^\infty(\mathbb{R}^s)$ with $\nabla\varphi_0 \in L^2(\mathbb{R}^s)$.

Proposition 6.1 [27] *Given a Hilbert space sector defined by a $\varphi_0 \in L^\infty(\mathbb{R}^s)$, a (renormalized) energy density can be defined on it with a convergent infinite volume integral if*

$$\nabla\varphi_0 \in L^2(\mathbb{R}^s). \tag{6.3}$$

Proof By Lemma 5.3 $G'(\chi)$ is globally Lipschitz continuous and therefore $G'(\chi) \in L^2(\mathbb{R}^s), \forall \chi \in H^1(\mathbb{R}^s)$. Now, from the identity

$$G(\chi_1) - G(\chi_2) = \int_0^1 d\sigma \frac{d}{d\sigma} G(\chi_1 + \sigma(\chi_2 - \chi_1))$$

$$= \int_0^1 d\sigma(\chi_2 - \chi_1)G'(\chi_1 + \sigma(\chi_2 - \chi_1)),$$

one has

$$\int d^s x |G(\chi_1) - G(\chi_2)| \leq \sup_{0 \leq \sigma \leq 1} \|G'(\chi_1 + \sigma(\chi_2 - \chi_1))\|_{L^2} \|\chi_2 - \chi_1\|_{L^2}$$

and, since $G(0) = 0$, $G(\chi) \in L^1(\mathbb{R}^s)$.

On the other hand,

$$\nabla\chi\nabla\varphi_0 + U'(\varphi_0)\chi = \nabla(\chi\nabla\varphi_0) - h(\varphi_0)\chi$$

---

[26] Ref. II (footnote 19).
[27] See Ref. II.

and the second term on the r.h.s. is integrable since $h \in H^{-1}(\mathbb{R}^s)$, $\chi \in H^1(\mathbb{R}^s)$. By eq. (6.3), $\chi \nabla \varphi_0 \in L^1(\mathbb{R}^s)$ and therefore the infinite volume limit of the integral of the first term vanishes. The other terms in eq. (6.1) are clearly integrable.

The condition (6.3) is actually necessary for the convergence of the infinite volume integral of the momentum density. Since, without loss of generality, we can take $\psi_0 = 0$, the background momentum subtraction vanishes and the *renormalized momentum density* is the conventional one

$$\mathcal{P}_{ren}(\varphi, \psi) = \psi \nabla \varphi \qquad (6.4)$$

Since ψ may be an arbitrary element of $L^2(\mathbb{R}^s)$, $\mathcal{P}_{ren}$ is integrable provided $\nabla \varphi \in L^2(\mathbb{R}^s)$, i.e. $\nabla \varphi_0 \in L^2(\mathbb{R}^s)$, since $\chi = \varphi - \varphi_0 \in H^1(\mathbb{R}^s)$.

It is worthwhile to remark that $\nabla \varphi_0 \in L^2$ implies in turn that $\nabla \varphi \in L^2(\mathbb{R})$, for all the elements of the corresponding HSS. A HSS defined by a $\varphi_0 \in L^\infty(\mathbb{R}^s)$ with $\nabla \varphi_0 \in L^2(\mathbb{R}^s)$ will be called a *Hilbert space sector with energy-momentum density*, or briefly a *physical sector*.

It is not difficult to show [28] that the infinite volume integrals of the renormalized energy-momentum densities define conserved quantities and that the corresponding functionals are continuous in the Hilbert space topology of the HSS.

A related question is the stability of a sector under external perturbations and an important rôle is played by the energy being bounded from below. Now, even if the potential is bounded from below, in general the renormalized energy may not be so. The renormalized energy is however bounded from below in the HSS sectors defined by absolute minima of the potential (vacuum sectors or phases) since $\nabla \varphi_0 = 0$ and $U(\varphi_0 + \chi) - U(\varphi_0) \geq 0$. The energy is not

---

[28] See Ref. II.

bounded from below in the sectors defined by relative minima of the potential (false vacuum sectors) and one expects instability against external field perturbations.

In conclusion, the set of solutions of the non-linear field equation (4.6), which have a reasonable physical interpretation are those belonging to Hilbert space sectors with energy-momentum density, (called *physical sectors*), and a distinguished rôle is played by the vacuum sectors or phases. (For time-independent solutions defining physical sectors see Appendix E). The analogy with the corresponding structures in quantum field theory [29] is rather remarkable.

---

[29] See references in footnotes 16 and 17.

7   An improved Noether theorem.
Spontaneous symmetry breaking

The existence of sectors, i.e. of "closed worlds" in the set of solutions of the non-linear equation (4.6), provides the mathematical and physical basis for the mechanism of spontaneous symmetry breaking, briefly discussed in Sect. 2. We can now understand the relation between the Noether theorem, the existence of conserved currents and the occurrence of spontaneous symmetry breaking which, among other things, imply the lack of existence of the corresponding charges.

To explain this we start with

Proposition 7.1 [30] *Let G denote the group of internal symmetries of eq. (4.6). Then*
1) G maps sectors into sectors and HSS into HSS

$$G : \mathcal{H}_\varphi \to \mathcal{H}_{g(\varphi)}, \ \ \forall g \in G,$$

giving rise to orbits of sectors and of HSS.
2) Each HSS $\mathcal{H}_\varphi$ determines a subgroup G_φ of G such that

$$G_\varphi : \mathcal{H}_\varphi \to \mathcal{H}_\varphi.$$

*G_φ is called the **stability group** of $\mathcal{H}_\varphi$ and $\mathcal{H}_\varphi$ is the carrier of a representation of its stability group*
3) A necessary and sufficient condition for G_φ being the stability group of $\mathcal{H}_\varphi$ is that there exists one element $\bar{\varphi} \in \mathcal{H}_\varphi$ such that

$$G_\varphi \bar{\varphi} \in \mathcal{H}_\varphi. \tag{7.1}$$

Furthermore, if $G_\varphi \bar{\varphi} = \bar{\varphi}$ and $\lambda_g = 1, \forall g \in G_\varphi$, then G_φ is represented by unitary operators in $\mathcal{H}_\varphi$.

---

[30]Ref. II.

Proof By the characterization of internal symmetries, eqs. (3.6), (3.7), $u'(t) - u(t) \in C^0(\mathbb{R}, H^1 \oplus L^2)$ implies

$$g(\varphi'(t)) - g(\varphi(t)) = A_g(\varphi'(t) - \varphi(t)) \in C^0(\mathbb{R}, H^1 \oplus L^2),$$

so that sectors are mapped into sectors.

Furthermore, if $u_0 = \{\varphi_0, \psi_0\}$ with $\varphi_0 \in L^\infty(\mathbb{R}^s)$, $\psi_0 \in L^2(\mathbb{R}^s)$ satisfies condition b) of Theorem 5.2, it follows that $A_g\varphi_0 + a_g \in L^\infty(\mathbb{R}^s)$, $A_g\psi_0 \in L^2(\mathbb{R}^s)$ and, by eqs. (3.6), (3.8),

$$\Delta g(\varphi_0) - U'(g(\varphi_0)) = A_g(\Delta\varphi_0 - U'(\varphi_0)) \in H^{-1}(\mathbb{R}^s),$$

i.e. g maps HSS into HSS.

Finally, for any element φ of $\mathcal{H}_{\varphi_0}$, putting $\chi = \varphi - \varphi_0$, one has

$$g(\varphi) - \varphi_0 = A_g\chi + g(\varphi_0) - \varphi_0 \tag{7.2}$$

and since for any $g \in G_{\varphi_0}$, $g(\varphi_0) - \varphi_0 \in H^1(\mathbb{R}^s) \oplus L^2(\mathbb{R}^s)$, by eq.(7.2) the mapping g induces an affine mapping on $H^1 \oplus L^2$ to which $\mathcal{H}_{\varphi_0}$ is naturally identified, by Theorem 5.2.

Conversely, by arguing as for eq.(7.2) if $\exists \bar{\varphi} \in \mathcal{H}_{\varphi_0}$ such that $g(\bar{\varphi}) - \bar{\varphi} \in H^1 \oplus L^2$ so does $g(\varphi) - \varphi_0$, i.e. $g \in G_\varphi$.

The other statements are obvious.

Since, as discussed before, different HHS define "disjoint physical worlds", an internal symmetry of the field equation (4.6) gives rise to a symmetry of the physical world described by the Hilbert sector $\mathcal{H}_\varphi$ only if it maps $\mathcal{H}_\varphi$ into $\mathcal{H}_\varphi$. Otherwise the symmetry is *spontaneously broken*. The notion of Hilbert space sectors then allows to recover and understand the concept of spontaneously broken symmetry at the level of classical field equations. As discussed in the Introduction, if $\mathcal{H}_\varphi$ is not stable under G, its elements cannot be classified in terms of irreducible representations of G.

It is now clear what distinguishes the infinite dimensional case with respect to the finite dimensional one. In the latter case, degenerate ground states related by a continuous symmetry, cannot be separated by potential barriers and one can move from one to the other by physically realizable operations. In the infinite dimensional case, degenerate ground states characterize different large distance behaviours of the field configurations, so that, even if they are related by a continuous symmetry, they cannot be related by physically realizable operations.

The crucial point is that for infinitely extended systems in order to connect two configurations of the systems one can dispose of operations which must both involve finite energy and be essentially localized.

When the field equations can be derived by a Lagrangean the link between the invariance group of the Lagrangean and the existence of conservation laws is provided by the classical Noether's theorem. However, the existence of a continuity equation or a local conservation law does not in general imply the existence of a constant of motion or *conserved charge*, since the integral which defines the charge,

$$Q^i = \int d^3x \, J_0^i(x)$$

may not converge.

Thus, the standard accounts of Noether theorem implicitly apply to the subset of solutions which decrease sufficiently fast at infinity, i.e. essentially to the "trivial vacuum" sector $\mathcal{H}_{\varphi=0}$. A criterium for the existence of a conserved charge implied by a continuity equation, in the general case when the solutions do not belong to $H^1 \oplus L^2$, is provided by the following improvement of Noether theorem.[31] Again the structure of Hilbert space sectors provides a simple solution of the problem.

For simplicity, we consider the case of real fields and of linear transformations ($a_g = 0$), the generalization being straightforward.

---

[31]F. Strocchi, loc.cit. (see footnote 23).

Theorem 7.2 *Let G be a N-parameter continuous (Lie) group of internal symmetries of the field equation (4.6) (or of the Lagrangean from which they are derived), then there exist N currents $J^i_\mu(u(x,t)) \equiv J^i_\mu(x,t)$, which obey the continuity equation*

$$\partial^\mu J^i_\mu(x,t) = 0, \qquad i = 1, ... N \qquad (7.3)$$

(**local conservation law**).

Given a physical HSS $\mathcal{H}_{\varphi_0}$, a one-parameter subgroup $G^{(j)} \subset G$ gives rise to a constant of motion or a **conserved Noether charge**

$$Q^j(u(t)) = Q^j(u(0)) \qquad (7.4)$$

$$Q^j(u(t)) \equiv \int d^s x \, J^j_0(u(x,t)) \qquad (7.5)$$

for all solutions $u(x,t) \in \mathcal{H}_{\varphi_0}$, iff $G^{(j)}$ is a subgroup of the stability group G_{φ_0} of $\mathcal{H}_{\varphi_0}$.

Proof We omit the proof of the first part, which is standard. For the second part, the stability of $\mathcal{H}_{\varphi_0}$ under $G^{(j)}$ is equivalent to its stability under infinitesimal transformations of $G^{(j)}$

$$\varphi \to \varphi + \epsilon^{(j)} \, \delta^{(j)}\varphi, \qquad \delta^{(j)}\varphi = \frac{\partial}{\partial \epsilon^{(j)}} A_{g_\epsilon}\varphi|_{\epsilon^{(j)}=0},$$

namely to the condition $\delta^{(j)}\varphi \in H^1(\mathbb{R}^s)$.

Now, $J^j_0(\varphi,\psi) = \psi \, \delta^{(j)}\varphi$ and therefore, since ψ may be an arbitrary element of $L^2(\mathbb{R}^s)$, $J^j_0 \in L^1(\mathbb{R}^s)$ iff $\delta^{(j)}\varphi \in L^2(\mathbb{R}^s)$. On the other hand, for a physical Hilbert space sector (see Sect. 6), $\nabla\varphi \in L^2(\mathbb{R}^s)$, which implies $\nabla A_g(\varphi) = A_g\nabla\varphi \in L^2(\mathbb{R}^s)$ and therefore $\nabla\delta^{(j)}\varphi = \delta^{(j)}\nabla\varphi \in L^2(\mathbb{R}^s)$. Hence, for a physical sector $\delta^{(j)}\varphi \in L^2(\mathbb{R}^s)$ is equivalent to $\delta^{(j)}\varphi \in H^1(\mathbb{R}^s)$.

Remark 1. It is not difficult to find the analog of the above Theorem in the more general case of non-internal symmetries, which commute with time evolution.

Remark 2. The notion of physical Hilbert space sector provides a neat characterization of the conditions for the existence of a conserved Noether charge, a point which seems to have been neglected in all the standard accounts of Noether theorem in classical field theory.[32] It is worthwhile to remark that a crucial point in the above theorem is the possibility of defining a conserved charge for *all* the elements of the HSS; the convergence of the integral (7.5) for some (but not all the) elements of $\mathcal{H}_{\varphi_0}$ is not enough for concluding that there is a conserved quantity, associated to the time evolution within the physical world defined by $\mathcal{H}_{\varphi_0}$. The impossibility of defining the generator Q^i of the symmetry on the whole $\mathcal{H}_{\varphi_0}$ is a sign that the symmetry is not realized in $\mathcal{H}_{\varphi_0}$.

In view of the above theorem, it is worthwhile to see more explicitly the possible mechanisms, by which the continuity equation for J^i_μ may fail to give rise to a conserved charge. To this purpose, one integrates $\partial^\mu J^i_\mu(x,t) = 0$ over the space-time region $\mathcal{V} \equiv \{x \in V = \text{a bounded space volume}, \ t \in [0,\tau]\}$ and uses Gauss theorem to get

$$0 = \int_\mathcal{V} d^s x \, dt \, \partial^\mu J^i_\mu(x,t) = Q^i_V(\tau) - Q^i_V(0) + \Phi_s(\vec{j}^{(i)}), \qquad (7.6)$$

where $\Phi_s(\vec{j})$ is the flux of $\vec{j}^{(i)} = \vec{\nabla}\varphi \, \delta^{(i)}\varphi$ over the space boundary $S \equiv \{x \in \partial V, \ t \in [0,\tau]\}$.

Now, to deduce the existence of a conserved charge one has to take the infinite volume limit $V \to \infty$ in eq. (7.6) and there are essentially two mechanisms by which the conservation law (7.4) may fail

1) in the limit $V \to \infty$, the flux $\Phi_s(\vec{j})$ vanishes, but Q_V does not converge. This is the case of the standard spontaneous symmetry

---

[32]See e. g. the references in footnote 9.

breaking and it is the strict analogue of the symmetry breaking a la Goldstone-Nambu. [33]

2) The flux $\Phi_S(\vec{j})$ does not vanish as $V \to \infty$; this is the analogue of the symmetry breaking induced by boundary effects or the symmetry breaking a la Higgs [34].

For physical HSS associated to the non-linear equation (4.6), the possibility 2) cannot arise since $\nabla\varphi \in L^2(\mathbb{R}^s)$ and, if the symmetry in question leaves the physical HSS stable, $\delta^{(i)}\varphi \in H^1(\mathbb{R}^s)$ so that $\nabla\varphi\delta^{(i)}\,\varphi \in L^1(\mathbb{R}^s)$ and the flux vanishes as $V \to \infty$.

A crucial rôle in the above analysis is played by the condition of finite energy-momentum, which in this case requires $\nabla\varphi_0 \in L^2(\mathbb{R}^s)$. This is no longer the case in gauge field theories, since the energy-momentum density involves the covariant derivative $(\nabla+A)\varphi$ (where A denotes the gauge potential), rather than $\nabla\varphi$. This opens the way to the Higgs mechanism of symmetry breaking for which the boundary effects give rise to a charge leaking at infinity [35].

[33] J. Goldstone, Nuovo Cim. **19**, 154 (1961); J. Goldstone, A. Salam and S. Weinberg, Phys. Rev. **127**, 965 (1962) Y. Nambu and G. Jona-Lasinio, Phys. Rev. **122**, 345 (1961); **124**, 246 (1961).

For a simple account see F. Strocchi, *Elements of Quantum Mechanics of Infinite Systems*, World Scientific 1985.

[34] P.W. Higgs, Phys. Lett. **12**, 132 (1964); T.W. Kibble, *Proc. Int. Conf. Elementary Particles*, Oxford, Oxford Univ. Press 1965; G.S. Guralnik, C.R. Hagen and T.W. Kibble, in *Advances in Particle Physics* Vol. 2, R.L. Cool and R.E. Marshak eds., Interscience New York 1968 and refs. therein. See also the references in the footnote below.

[35] G. Morchio and F. Strocchi, in *Fundamental Problems of Gauge Field Theory*, G. Velo and A. S. Wightman eds. Plenum 1986; F. Strocchi, in *Fundamental Aspects of Quantum Theory*, V. Gorini and A. Frigerio eds., Plenum 1986.

8 Examples

1) Non-linear scalar field in one space dimension

The model describes the simplest non-linear field theory and it can be regarded as a prototype of field theories in one space dimension ($s = 1$). The model can also be interpreted as a non-linear generalization of the wave equation. The interest of the model is that, even at the classical level, it has stable solutions with a possible particle interpretation [36].

The model is defined by the potential

$$U = -\tfrac{1}{2}m^2\varphi^2 + \tfrac{1}{4}\lambda\varphi^4 = \frac{1}{4}\lambda(\varphi^2 - \frac{m^2}{\lambda})^2 - \frac{1}{4}\frac{m^4}{\lambda} \qquad (8.1)$$

and therefore the equations of motion read

$$\Box\varphi = -\lambda\varphi(\varphi^2 - \mu^2), \quad \mu^2 \equiv m^2/\lambda. \qquad (8.2)$$

i) *Vacuum state solutions* The simplest solutions are the *ground state solutions*, invariant under space and time translations, i.e. $\varphi = const.$ If the field φ takes values in $\mathbb{R}$, there are only three possibilities

$$\varphi_0^{\pm} = \pm\mu, \qquad \varphi_0 = 0. \qquad (8.3)$$

By the discussion of Sects. 5-7, $\varphi_0^{\pm}$ define disjoint Hilbert space sectors $\mathcal{H}_{\pm}$, for which an energy-momentum density can be defined and for which the energy is bounded below. The other constant solution $\varphi_0 = 0$, corresponding to the so-called trivial vacuum sector, still defines a Hilbert space sector with energy-momentum density, but the energy is not bounded below and therefore in this case the sector is not energetically stable under external perturbations (see

---

[36]J. Goldstone and R. Jackiw, Phys. Rev. **D11**, 1486 (1975). See also R. Rajaraman, *Solitons and Instantons*, North-Holland 1982 and references therein.

Sect. 7). This would be the only vacuum state solution available in Segal's approach.

If the field φ takes values in $\mathbb{R}^n, n > 1$, the internal symmetry group is the continuous group G of transformations (3.6), (3.7) with $\lambda = 1, a = 0$. In this case, besides the trivial vacuum solution $\varphi_0 = 0$, the non-trivial vacuum solutions are given by the points of the orbit

$$\{\varphi_0^g \equiv A_g\bar{\varphi}_0, \quad g \in G, \quad \bar{\varphi}_0^2 = \mu^2\}. \tag{8.4}$$

For $n = 1$, the internal symmetry group is the discrete group

$$Z_2 : \varphi \rightarrow -\varphi.$$

Clearly, in all cases, the internal symmetry group is unbroken in the trivial vacuum sector $\mathcal{H}_0$, but it is spontaneously broken in each "pure phase" $\mathcal{H}_g$, defined by φ_0^g.

ii) *Time independent solutions defining physical Hilbert space sectors. Kinks*

Another interesting class are the time independent solutions, which satisfy

$$(\frac{d}{dx})^2\varphi = \lambda\varphi(\varphi^2 - \mu^2). \tag{8.5}$$

This implies

$$\frac{d}{dx}(\tfrac{1}{2}\varphi_x^2 - \tfrac{1}{4}\lambda(\varphi^2 - \mu^2)^2) = 0, \quad \varphi_x \equiv d\varphi/dx,$$

i.e.

$$\tfrac{1}{2}\varphi_x^2 = \tfrac{1}{4}\lambda(\varphi^2 - \mu^2)^2 + C, \quad C = constant. \tag{8.6}$$

For simplicity, we consider the case in which φ takes values in $\mathbb{R}$, leaving the straightforward generalization as an exercise.

The discussion of the solutions of eq. (8.5), as given in the literature, (see e.g. the references in the previous footnote), is done under the condition that they have finite energy when the potential is so renormalized that it vanishes at its absolute minimum. This means that

$$\tfrac{1}{2}(\nabla\varphi)^2 + \tfrac{1}{4}\lambda(\varphi^2 - \mu^2)^2 \in L^1.$$

By the discussion of Sect. 1.5, this appears as too restrictive, since it does not consider the possibility of energy renormalization, (eq. 6.2), and in particular it crucially depends on the overall scale of the potential (it also excludes the trivial vacuum solution $\varphi_0 = 0$!). For these reasons we prefer to leave open the energy renormalization.

To simplify the discussion we will only assume that φ has (bounded) limits $\varphi(\pm\infty)$, when $x \to \pm\infty$ (regularity at infinity). Then, quite generally, since U is by assumption of class C^2, also $U'(\varphi)$ has bounded limits as $x \to \pm\infty$ and eq.(8.5) implies that so does also $d^2\varphi/dx^2$. On the other hand, for any test function f of compact support, with $\int f(x)dx = 1$,

$$\lim_{a\to\pm\infty} (\Delta\varphi)(x+a) = \lim_{a\to\pm\infty} \int \Delta\varphi(x+a)\ f(x)dx$$

$$= \lim_{a\to\pm\infty} \int \varphi(x+a)\Delta f(x)dx = \varphi(\pm\infty) \int \Delta f(x)\ dx = 0.$$

Then, eq. (8.5) implies

$$U'(\varphi(\pm\infty)) = 0. \tag{8.7}$$

Now, for physical sectors $\nabla\varphi \in L^2$, so that the constant C in eq. (8.6) must vanish and one has

$$\varphi(x) = \varepsilon(x)\sqrt{\tfrac{\lambda}{2}}(\varphi^2 - \mu^2), \tag{8.8}$$

with $\varepsilon(x)^2 = 1$. Actually, eq. (8.5) implies that $\varepsilon(x)$ is independent of x, i.e. $\varepsilon(x) = \pm 1$. Eq. (8.8) can be easily integrated and it gives

$$\varphi_x = \mp\mu\tanh(\sqrt{\tfrac{\lambda}{2}}\mu(x - a)), \tag{8.9}$$

where a is an integration constant.

The plus/minus sign gives the so-called *kink/anti-kink* solution, respectively. Such solutions do not vanish at $x \to \pm\infty$, but, nevertheless, they have some kind of localization, since they significantly differ from the constants φ_0^+, φ_0^- only in a region of width $(\sqrt{\lambda}\mu)^{-1}$. They are not local perturbations of the ground state solutions $\varphi_0^\pm$ and in fact they define different Hilbert sectors $\mathcal{H}_k, \mathcal{H}_{\bar{k}}$. The corresponding renormalized energy momentum density is defined by

$$\mathcal{E}_{ren} = \tfrac{1}{2}(\nabla\varphi)^2 + U(\varphi) + \tfrac{1}{4}\lambda\mu^4 = \tfrac{1}{2}(\nabla\varphi)^2 + \tfrac{1}{4}\lambda(\varphi^2 - \mu^2)^2$$

and it is localized around the "centre of mass" of the kink, namely $x = a$. (It is instructive to draw the shape of the kink solution). The total renormalized energy is

$$E_k = \frac{2\sqrt{2}}{3}\frac{m^3}{\lambda} \tag{8.10}$$

and it clearly exhibits the non-perturbative nature of the kink solution.

iii) *Moving kink. Particle behaviour*

Since eq. (8..2) is invariant under a Lorentz transformation

$$x \to x' = (x - vt)/\sqrt{1 - v^2}, \quad t \to t' = (t - vx)/\sqrt{1 - v^2},$$

(where the velocity of light c is put $= 1$) if $\varphi(x, t)$ is a solution, so is also $\varphi'(x, t) \equiv \varphi(x', t')$. Thus, from the static solutions (8.9), we can generate time dependent ones (for simplicity we put $a = 0$)

$$\varphi(x, t) = \mp\mu\tanh(\sqrt{\tfrac{\lambda}{2}}\mu(x - vt)/\sqrt{1 - v^2}), v^2 < 1. \tag{8.11}$$

The energy-momentum density is localized around the point $x = vt$ ("center of mass" of the kink), which moves with velocity v (*moving kink* solution).

Clearly, $\varphi(x, t) - \varphi(x, 0) \in C^0(\mathbb{R}, H^1)$, i.e. $\varphi(x, t)$ defines a sector. Furthermore $\varphi(x, 0) \in L^\infty(\mathbb{R})$, $\psi(x, 0) = \dot\varphi(x, 0) \in L^2(\mathbb{R})$ and obviously condition b) of Theorem 5.1 is satisfied; then $(\varphi(x, 0), \psi(x, 0))$ defines a Hilbert space sector.

This implies the stability of such solutions under $H^1 \oplus L^2$ perturbations (see Sect. 5). This settles the problem of stability of the kink sector [37] and, thanks to Theorem 5.1, the proof does not involve expansions or linearizations. It is not difficult to see that the static kink solution, corresponding to $v = 0$ in eq. (8.11), belongs to the same sector defined by the corresponding moving kink solution.

From a physical point of view (energy-momentum localization and stability), the kink is a candidate to describe particle-like excitations associated to eq. (8.2) and in fact, in the past, this feature has motivated attempts to use such kink-like solution as a non-perturbative semi-classical approach to the descriptions of baryons in quantum field theory [38].

---

[37]See e.g. R. Rajaraman, Phys. Rep. **21**, 227 (1975), especially Sect. 3.2.

[38]R.F. Dashen, B. Hasslacher and A. Neveu, Phys. Rev. **D10**, 4130 (1974); J. Goldstone and R. Jackiw, Phys. Rev. **D11**, 1486 (1975); for a rich collection of important papers see C. Rebbi and G. Soliani, *Solitons and Particles*, World Scientific 1984.

2) The Sine-Gordon equation

The Sine-Gordon equation is

$$\Box \varphi = -g \sin \varphi, \tag{8.12}$$

where $\varphi(x, t)$ is a scalar field in one space dimension. It is of great interest in various fields of theoretical physics, like propagation of crystal dislocation, magnetic flux in Josephson lines, Bloch wall motion in magnetic crystals, fermion bosonization in the Thirring model of elementary particle interactions [39], etc.

i) *Static solutions*

The simplest static solutions are the constants

$$\varphi = \pi n, \quad n \in \mathbb{Z}. \tag{8.13}$$

They all define disjoint Hilbert space sectors and for n even correspond to absolute minima of the potential

$$U = g(1 - \cos \varphi). \tag{8.14}$$

In this case the energy is bounded below in the corresponding Hilbert sectors.

The internal symmetries of eq. (8.12) are

$$\varphi \to \varphi + 2\pi n$$

and

$$\varphi \to -\varphi$$

---

[39]See A. Barone, F. Esposito and C. J. Magee, Theory and Applications of the Sine-Gordon Equation, in Riv. Nuovo Cim. **1**, 227 (1971); A.C. Scott, F.Y. Chiu, and D.W. Mclaughlin, Proc.I.E.E.E. **61**, 1443 (1973); G.B. Whitham, *Linear and Non-Linear Waves*, J. Wiley 1974; S. Coleman, Phys. Rev. **D11**, 2088 (1975); S. Coleman, *Aspects of Symmetry*, Cambridge Univ. Press 1985; J. Fröhlich, in *Invariant Wave Equations*, G. Velo and A. S. Wightman eds., Springer-Verlag 1977.

They are broken in the sectors $\mathcal{H}_{\pi n}$ defined by the vacuum solutions (8.13).

To determine other non trivial static solutions we proceed as in Example 1). They obey the equation

$$\Delta\varphi = g\sin\varphi, \qquad (8.15)$$

which implies

$$\frac{d}{dx}[\tfrac{1}{2}\varphi_x^2 + g\cos\varphi] = 0, \qquad (8.16)$$

i.e.

$$\tfrac{1}{2}\varphi_x^2 + g\cos\varphi = C, \qquad C = constant. \qquad (8.17)$$

As in the previous example, we prefer to leave open the energy renormalization and we classify all the solutions of eq. (8.17) which have (bounded) limits $\varphi_{\pm\infty}$ when $x \to \pm\infty$. By the same argument as before, one finds that $\sin\varphi_{\pm\infty} = 0$, i.e.

$$\varphi_{\pm\infty} = \pi n_\pm, \qquad n_\pm \in \mathbb{Z} \qquad (8.18)$$

and, from the condition $\nabla\varphi \in L^2$, one gets

$$n_+ = n_- \bmod 2\pi, \qquad C = \varepsilon g,$$

with $\varepsilon = 1$ for $n_+ = even$, $\varepsilon = -1$ for $n_+ = odd$. Actually, the case $n_+ = odd$ is ruled out by eq. (8.17), which requires $C - g\cos\varphi = \tfrac{1}{2}\varphi_x^2 \geq 0$. Then

$$\varphi_x = \varepsilon(x)\sqrt{2g}\sqrt{1-\cos\varphi}, \qquad \varepsilon(x)^2 = 1 \qquad (8.19)$$

and again $\varepsilon(x) = \pm1$, by eq. (8.16).

Eq. (8.17) can be easily integrated and it gives

$$\varphi(x) = \pm4\tan^{-1}[\exp\sqrt{g}(x-a)] \equiv \varphi_{s/\bar{s}} \qquad (8.20)$$

with a an integration constant. Corresponding to the $+$ or $-$ sign the solution is called *soliton* or *anti-soliton*.

As before, moving soliton (or anti-soliton) solutions can be obtained by Lorentz transformations, i.e. by replacing $x - a$ in eq. (8.20) by $(x - a - vt)/\sqrt{1 - v^2}$. A remarkable property of solitons with respect to kinks is that they are unaltered by scattering. The literature on solitons is vast (see e.g. the references in the previous footnote).

It is not difficult to see that φ_s and $\varphi_{\bar{s}}$ define different Hilbert sectors $\mathcal{H}_s, \mathcal{H}_{\bar{s}}$ (also different from the $\mathcal{H}_{\pi n}$, defined by the vacuum solution (8.13)).

9   The Goldstone theorem

The mechanism of SSB does not only provide a general strategy for unifying the description of apparently different systems, but it also provide information on the energy spectrum of an infinite dimensional system, by means of the so-called *Goldstone theorem*, [40] according to which to each broken generator T of a continuous symmetry there corresponds a massless mode, i.e. a free wave. The quantum version of such a statement has been turned into a theorem, [41] whereas, as far as we know, no analogous theorem has been proved for classical (infinite dimensional) systems and the standard accounts rely on heuristic arguments.

The standard heuristic argument, which actually goes back to Goldstone, considers as a prototype the nonlinear equation (3.1)

$$\Box\varphi + U'(\varphi) = 0,$$

where the multi-component real field φ transforms as a linear representation of a Lie group G and the potential U is invariant under the transformations of G. This implies that for the generator T^α one has

$$0 = \delta^\alpha U(\varphi) = U'_j(\varphi)\, T^\alpha_{jk}\, \varphi_k, \quad \forall \varphi \tag{9.1}$$

and therefore the derivative of this equation at $\varphi = \overline{\varphi}$ gives

$$U''_{jk}\, (T^\alpha\overline{\varphi})_k = 0. \tag{9.2}$$

Thus, in an expansion of the potential around $\overline{\varphi}$, the quadratic term, which has the meaning of a mass term, has a zero eigenvalue in the direction $T^\alpha\overline{\varphi}$. This is taken as evidence that there is a massless mode. In our opinion, the argument is not conclusive since it involves an expansion and one should in some way control the effect of

---

[40] J. Goldstone, Nuovo Cimento **19**, 154 (1961)

[41] J. Goldstone, A. Salam and S. Weinberg, Phys. Rev. **127**, 965 (1962); J. Swieca, *Goldstone's theorem and related topics*, Cargese lectures 1969

higher order terms; moreover, it is not clear that there are (physically meaningful) solutions in the direction of $T^\alpha \overline\varphi$ for all times, so that for them the quadratic term disappears. In any case, the argument does not show that there are massless solutions as in the quantum case.

Another heuristic argument appeals to the finite dimensional analogy, where the motion of a particle along the bottom of the potential, i.e. along the orbit $\{g^\alpha(\lambda)\overline\varphi\}$, where $g^\alpha(\lambda)$, $\lambda \in \mathbf{R}$, is the one parameter subgroup generated by T^α, does not feel the potential, since $U'(g^\alpha\overline\varphi)=0$, and therefore the motion is like a free motion. This is considered as evidence that, correspondingly, in the infinite dimensional case there are massless modes. Again the argument does not appear complete, since it is not at all clear that there are physically meaningful solutions, i.e. belonging to the physical sector of $\overline\varphi$ and therefore of the form $\varphi=\overline\varphi+\chi$, $\chi \in H^1(\mathbf{R}^s)$, s = space dimension, of zero mass.

We propose a version[42] of the Goldstone theorem for classical fields as a mathematically acceptable substitute and correction of the above heuristic arguments . We consider the case of space dimension $s = 3$, unless otherwise stated and for simplicity the case of compact semi-simple Lie group G of internal symmetries. The potential is assumed to be of class C^3.

We first recall that, given a solution $u(t)$ of the integral equation (4.6), its asymptotic time $(t \to \pm\infty)$ behaviour define the so-called *scattering configurations* or *asymptotic states* $u_\pm(t)$ associated to $u(t)$. The behaviour of $f(u)$ near $u = 0$ plays a crucial rôle for such asymptotic limits and if $f(u)-f'(0)u$ vanishes to a sufficiently high degree, e.g. as $O(u^3)$, such limits $u_\pm(t)$ exist and their time evolution is that corresponding to the differential operator $\Box + f'(0)$, i.e.

$$u_\pm(t') = \mathcal{W}(t' - t)\, u_\pm(t),$$

_____

[42]F. Strocchi, Phys. Lett. **A267**, 40 (2000).

where $\mathcal{W}(t)$ is the propagator corresponding to the differential operator $\square + f'(0)$, (if $f'(0) = 0$, $\mathcal{W}(t)$ is the free wave equation propagator
$W(t)$ defined in Sect.4).

The mathematical theory of scattering for the nonlinear wave
equation is well developed and it is beautifully reviewed by W.
Strauss, *Non-linear Wave Equations*, Am. Math. Soc. 1989.

The mathematical problem of the existence of the scattering configurations (the so-called scattering theory) is to guarantee the well
definiteness of the Yang-Feldman equations

$$u_{\pm}(t) = u(t) + \int_{t}^{\pm\infty} ds\, U_0(t-s)\, f(u(s)), \qquad (9.3)$$

which express $u_{\pm}(t)$ in terms of the solution $u(t)$ and of the propagator $\mathcal{W}$. The Yang-Feldman equations can be interpreted as a form of
the integral eq.(4.6) with initial data given at $t = \pm\infty$, respectively.
The problem of the existence of the asymptotic limits reduces to estimating the asymptotic time decay of the nonlinear term $f(u(s))$
such that the integrals on the r.h.s. of the Yang-Feldman equations
exist. This can be done by using the Basic L^∞ estimates on the time
decay of the free solutions (see Strauss' book pp. 5-6).

For small amplitude solutions, i.e. for initial data small in some
norm, e.g. of the form εu for fixed u, the asymptotic limits are
completely governed by the behaviour of $f(u)$ near $u = 0$.

Theorem 9.1 *Let G be an N-parameter continuous (Lie) group of
internal symmetries of the nonlinear equation (3.1) and $\mathcal{H}_{\overline{\varphi}}$ a Hilbert
Space Sector (HSS), defined by the absolute minimum $\overline{\varphi}$ of the potential U, where G is spontaneously broken down to $G_{\overline{\varphi}}$, the stability
group of $\overline{\varphi}$.*

Then, for any generator T^α, such that $T^\alpha\overline{\varphi} \neq 0$,

*i)there are scattering configurations, associated to solutions belong-
ing to the sector $\mathcal{H}_{\overline{\varphi}}$, which are solutions of the free wave equation*
(Goldstone modes).

*ii) for any sphere Ω_R of radius R and any time T there are solutions $\varphi_G^\alpha(x,t) \neq \overline{\varphi}$, $\varphi_G^\alpha \in \mathcal{H}_{\overline{\varphi}}$, whose propagation in Ω_R in the time interval $t \in [0,T]$ is that of free waves (**Goldstone-like solutions**).*

Proof i) For solutions $\phi \in \mathcal{H}_{\overline{\varphi}}$, i.e. of the form $\varphi = \overline{\varphi} + \chi$, $\chi \in H^1$, the conservation of the current $j_\mu = (\partial_\mu \phi) T^\alpha \phi$, associated to the generator T^α, (without loss of generality we can take φ real and T^α antisymmetric), reads

$$0 = \partial_\mu j^\mu = \Box \chi_i T^\alpha_{ij} \phi_j = \Box(\chi_i T^\alpha_{ij} \overline{\varphi}) + \Box \chi_i T^\alpha_{ij} \chi_j, \qquad (9.4)$$

and by the invariance of the potential, the second term can be written as $-U'(\overline{\varphi} + \chi)_i T^\alpha_{ij} \overline{\varphi}_j$. (In the quantum case, thanks to the vacuum expectation value, one has only the analogue of the first term and the proof gets simpler).

Now, for small amplitude solutions χ, the asymptotic limits are governed by the behaviour of $\mathcal{U}'(\chi) \equiv U'(\overline{\varphi} + \chi)$ near $\chi = 0$ and in this region, by the invariance of the potential, one has

$$\mathcal{U}'_i(\chi)\,(T^\alpha \overline{\varphi})_i = U'''_{ijk}(\overline{\varphi})\,\chi_j \chi_k\,(T^\alpha \overline{\varphi})_i + O(\chi^3).$$

This implies that the small amplitude mode $\chi^\alpha \equiv \chi_i\,(T^\alpha \overline{\varphi})_i$ satisfies a nonlinear wave equation with an effective potential which vanishes to a degree $p \geq 3$ near $\chi = 0$. Thus, the large time decay of the nonlinear term appearing in the corresponding Yang-Feldman equation, is not worse than in the case of a wave equation with potential vanishing with degree $p \geq 3$, near the origin, (other massive modes occurring in $\mathcal{U}'$ have faster decay properties). Then, one can appeal to standard results [43] to obtain the existence of the asymptotic limits $\chi_\pm^\alpha(t)$ satisfying the free wave equation.

ii) The existence of free waves $\varphi(x,t) = \overline{\varphi} + \chi(x,t)$ within a given

---

[43]H. Pecher. Math. Zeit. **185**, 261 (1984); **198**, 277 (1988).

region Ω_R in the time interval $[0, T]$ is equivalent to $U'(\overline{\varphi} + \chi(x, t)) = 0$, $\forall x \in \Omega_R$, $t \in [0, T]$, so that if the absolute minima of the potential consist of a single orbit $\overline{\varphi} + \chi(x, t) = \exp\left(h^\alpha(x, t) T^\alpha\right) \overline{\varphi}$, $h^\alpha(x, t)$ real $\in H^1$ and for solutions associated to a given generator T^α, with $T^\alpha \overline{\varphi} \neq 0$, one has solutions of the form

$$\varphi^\alpha(x, t) = e^{h(x,t) T^\alpha} \overline{\varphi}.$$

Now, the wave equation $\Box \varphi(x, t) = 0$, requires

$$\Box h(x, t) = 0, \quad (\partial_\mu h \, \partial^\mu h)(x, t) = 0, \tag{9.5}$$

(since T^α and $(T^\alpha)^2$ have different symmetry properties).

This implies that any C^2 function of h also satisfies eqs.(7) and in particular so does $\chi^{(\alpha)} \equiv \varphi^{(\alpha)} - \overline{\varphi}$.

Equations (9.5) have solutions of the form $\chi(x, t) = h_k(x, t) = h(k_0 t - \mathbf{k} \cdot \mathbf{x})$, with h an arbitrary C^2 function and $k = (k_0, \mathbf{k})$ a light-like four vector, but they are not in $H^1(\mathbf{R}^s)$ for $s \geq 2$. One can argue more generally that the above equations do not have solutions $h \in H^1(\mathbf{R}^s)$ for $s \geq 2$. In fact, the wave equation requires that the support of the $s + 1$-dimensional Fourier transform $\hat{h}(k)$, $k \in \mathbf{R}^{s+1}$ is contained in $\{k^2 = 0\}$, and the second equation becomes

$$k^2 \int d^{s+1} q \, \overline{\hat{h}}(q - k) \, \hat{h}(q) = 0,$$

since $kq - q^2 = k^2 - (k - q)^2$, $(k - q)^2 \hat{h}(k - q) = 0$. Thus

$$H(k) \equiv \int d^{s+1} q \, \overline{\hat{h}}(q - k) \, \hat{h}(q)$$

must have support in $k^2 = 0$. Now, the sum of two light-like four vectors $k - q$, q may be a light-like vector k only if $\mathbf{k}$ and $\mathbf{q}$ are parallel or antiparallel, corresponding to $\mathrm{sign} k_0 q_0 = +1$ or $= -1$, respectively, i.e. only if $q = \lambda k$, $\lambda \in \mathbf{R}$. Hence, if $k \in \mathrm{supp} H$ and q and $q - k$ belong to the support of $\hat{h}$, q must lie in the intersection of the light cone $q^2 = 0$ and the hyperplane $kq = 0$; thus, writing

$$\hat{h}(q) = \delta(q^2) \, h_r(\mathbf{q}), \quad H(k) = \delta(k^2) \, H_r(\mathbf{k}),$$

where δ denotes the Dirac delta function, one has

$$H_r(\mathbf{k}) = \mu(I_k) \int d\lambda\, h_r(\mathbf{k}(1-\lambda))\, h_r(\lambda \mathbf{k}),$$

where μ is the Lebesgue measure and

$$I_k \equiv \{\mathbf{q}; \mathbf{q} \in \mathrm{supp} h_r \cap \{kq = 0, k^2 = 0, q^2 = 0\}\}$$

For $s \geq 2$ this appears to exclude that $h \in H^1(\mathbf{R}^s)$.

The above argument indicates that the solutions with the properties of ii) can be constructed, e.g. as

$$\varphi_G^\alpha(x, t) = e^{h_k(x,t)\, f_{R+2T}(\mathbf{x})\, T^\alpha}\, \overline{\varphi},$$

with $f_R(\mathbf{x}) = 1$ for $|\mathbf{x}| \leq R$ and $= 0$ for $|\mathbf{x}| \geq R(1+\varepsilon)$.

The above discussion also shows that in one space dimension $s = 1$ one may find solutions of eqs.(9.5) belonging to H^1 and therefore prove the existence of genuine Goldstone modes all over the space. In fact, any function $h(x-t)$ or $h(x+t)$, $h \in H^1(\mathbf{R})$, is a solution of eqs.(9.5).

Appendices

A   Properties of the free wave propagator

a) $W(t)$ *maps* $\mathcal{S} \times \mathcal{S}$ *into* $\mathcal{S} \times \mathcal{S}$

If $u \in \mathcal{S}(\mathbb{R}^s) \times \mathcal{S}(\mathbb{R}^s)$ ($\mathcal{S}(\mathbb{R}^s)$ is the Schwartz space of C^∞ test functions decreasing at infinity faster than any inverse polynomial), then the solution of the free wave equation is easily obtained by Fourier transform and one has

$$W(t) \begin{pmatrix} \varphi_0(k) \\ \psi_0(k) \end{pmatrix} = \begin{pmatrix} \cos|k|t & (\sin|k|t)/|k| \\ -|k|\sin|k|t & \cos|k|t \end{pmatrix} \begin{pmatrix} \varphi_0(k) \\ \psi_0(k) \end{pmatrix}. \qquad (A.1)$$

It is easy to check that $\cos|k|t$, $(\sin|k|t)/|k|$ etc. are multipliers of $\mathcal{S}$, i.e. they map $\mathcal{S}$ into $\mathcal{S}$ and are continuous in t. It is also immediate to check that

$$\frac{d}{dt}W(t)\big|_{t=0} = \begin{pmatrix} 0 & 1 \\ |k|^2 & 0 \end{pmatrix} = K \qquad (A.2)$$

and that the group property holds.

b)*Hyperbolic character of* $W(t)$. *Huygens' principle.*

Let Ω_R be a sphere of radius R in $\mathbb{R}^s$, which for simplicity we take centered at the origin, and Ω_{R-t} the concentric sphere of radius

$R - t, \ 0 \le t \le R - \delta, \ \delta > 0$, then

$$\|W(t) \, u_0\|_{\Omega_{R-t}} \le e^{|t|/2} \, \|u_0\|_{\Omega_R}, \qquad (A.3)$$

the norms being those defined by eq.(4.8).

Eq. (A.3) means that the norm of $u(t)$ in Ω_{R-t} depends only on the norm of $u(0)$ in Ω_R (influence domain) and it is a mathematical formulation of Huygens' principle. We start by proving (A.3) for $u \in \mathcal{S} \times \mathcal{S}$. The free wave equation implies the following equation of energy-momentum conservation

$$\tfrac{1}{2}\frac{d}{dt}[(\nabla\varphi)^2 + \psi^2] - \vec{\nabla} \cdot (\psi\vec{\nabla}\varphi) = 0 \qquad (A.4)$$

and, by adding to both sides $\varphi\psi = d(\tfrac{1}{2}\varphi^2)/dt$, one has

$$\tfrac{1}{2}\frac{d}{dt}[(\nabla\varphi)^2 + \varphi^2 + \psi^2] - \vec{\nabla}(\psi\vec{\nabla}\varphi) = \varphi\psi. \qquad (A.5)$$

Now, we integrate the above equation over the cut cone with lower base Ω_R and upper base Ω_{R-t}, and we use Gauss' theorem to transform the volume integral into a surface integral and we get

$$\|u(t)\|_{\Omega_{R-t}}^2 - \|u(0)\|_{\Omega_R}^2 + \int_S dS \, \{\tfrac{1}{2}[(\nabla\varphi)^2 + \varphi^2 + \psi^2]n_0 - \vec{n} \cdot (\psi\vec{\nabla}\varphi)\}$$

$$= \int_0^t d\tau \int_{\Omega_{R-\tau}} \varphi(x,\tau)\psi(x,\tau) \, d^s x, \qquad (A.6)$$

where S is the three-dimensional surface defined by $|x| = R - \tau$, $0 \le \tau \le t$ and $n = (\vec{n}, n_0)$ is its outer normal. Since $n_0 > 0$ and $|\vec{n}| = n_0$ we have that the function in curly brackets in eq. (A.6) is greater than

$$n_0 \tfrac{1}{2}[(\nabla\varphi)^2 + \varphi^2 + \psi^2 - 2|\psi| \, |\nabla\varphi|] \ge 0.$$

Furthermore, by the inequality $a^2 + b^2 \ge 2ab$ the integral over $\Omega_{R-\tau}$ on the r.h.s. of eq. (A. 6), is majorized by $\|u(\tau)\|_{\Omega_{R-\tau}}^2$. Hence we get

$$\|u(t)\|_{R-t}^2 \le \|u(0)\|_{\Omega_R}^2 + \int_0^t d\tau \|u(\tau)\|_{\Omega_{R-\tau}}^2. \qquad (A.7)$$

Now, by *Gronwall's lemma* [1] if a non-negative continuous function $F(t)$ satisfies

$$F(t) \leq A(t) + \int_0^t d\tau \, B(\tau) F(\tau), \qquad (A.8)$$

with $A(t), B(t)$ both continuous and non-negative and $A(t)$ non decreasing, then

$$F(t) \leq A(t) \exp\left(\int_0^t B(\tau) d\tau\right). \qquad (A.9)$$

By using Gronwall's lemma, eq. $(A.7)$ yields eq. $(A.3)$.

Eq. $(A.3)$ also implies that $W(t)$ is a continuous operator with respect to the X_{loc} topology and then it can be extended from the dense domain $\mathcal{S} \times \mathcal{S}$ to whole X_{loc} preserving eq. $(A.3)$ and the group law.

c) $W(t)$ *is a strongly continuous group*

Thanks to the group law, it is enough to show the strong continuity at $t = 0$, namely that, for any bounded region V in $\mathbb{R}^S$,

$$\lim_{t \to 0} \|(W(t) - 1)u\|_V = 0, \qquad \forall u \in X_{loc}. \qquad (A.10)$$

Eq. $(A.10)$ is obvious for $u \in \mathcal{S} \times \mathcal{S}$ (see eq. $(A.1)$) and it can be extended to the whole X_{loc} by using eq. $(A.3)$. In fact if $u_j \in \mathcal{S} \times \mathcal{S}$ and $u_j \to u$ in X_{loc}, one has

$$\|(W(t) - 1)u\|_V \leq \|(W(t) - 1)u_j\|_V + \|(W(t) - 1)(u_j - u)\|_V$$

and the latter term is majorized by $(e^{\frac{1}{2}} + 1)\|u_j - u\|_{\Omega_R}$, where Ω_R is a sphere such that $\Omega_{R-1} \supset V$, as a consequence of eq. $(A.3)$.

One can show that the domain of the generator K is $H^2_{loc}(\mathbb{R}^S) \oplus H^1_{loc}(\mathbb{R}^S)$; in fact, $\forall u \in X_{loc} \subset \mathcal{S}' \times \mathcal{S}'$, in the distributional sense

---

[1] See e.g. G. Sansone and R. Conti, *Non-linear Differential Equations*, Pergamon Press 1964, p.11.

from eq. (A.2) one has

$$K \begin{pmatrix} \varphi \\ \psi \end{pmatrix} = \begin{pmatrix} \psi \\ \Delta\varphi \end{pmatrix}.$$

The condition that the r.h.s. belongs to X_{loc}, gives $\psi \in H^1_{loc}(\mathbb{R}^S)$ and $\Delta\varphi \in L^2_{loc}(\mathbb{R}^S)$, which is equivalent to $\varphi \in H^2_{loc}(\mathbb{R}^S)$.

B   The Cauchy problem for small times

Theorem B.1 *If $f(u)$ satisfies a local Lipschitz condition then properties 1), 2), 3), 4) listed in Sect. 4, hold*

Proof.[2] 1) One has to check that $W(t - s)f(u(s))$ is an integrable function; it is enough to show that it is a continuous function in the X_{loc} topology. To this purpose, we consider the inequality ($f_s \equiv f(u(s))$)

$$\|W(t-s)f_s - W(t-s')f_{s'}\|_{\Omega_{R-t}} \leq \qquad\qquad (B.1)$$
$$\leq \|(W(t-s) - W(t-s'))f_s\|_{\Omega_{R-t}} + \|W(t-s')(f_s - f_{s'})\|_{\Omega_{R-t}}$$

The first term on the right hand side goes to zero as $s' \to s$ as a consequence of the strong continuity of $W(t)$ on X_{loc} (see Appendix A, c)). The second term can be estimated by using the hyperbolic character of $W(t)$

$$\|W(t)u_0\|_{\Omega_{R-t}} \leq e^{|t|/2}\|u_0\|_{\Omega_R}$$

(see Appendix A, b)) and the local Lipschitz property of f

$$\|W(t-s')(f_{s'} - f_s)\|_{\Omega_{R-t}} \leq Ae^{|t-s'|/2}\|f_{s'} - f_s\|_{\Omega_R} \leq$$
$$\leq Ae^{|t-s'|/2}\|u(s') - u(s)\|_{\Omega_R}$$

The r.h.s. goes to zero as $s' \to s$, if $u(t)$ is continuous in time.

2), 3) For any two solutions $u_1(t), u_2(t)$, continuous in time, one has by the hyperbolic character of the free wave equation and the local Lipschitz property

$$\|u_1(t) - u_2(t)\|_{\Omega_{R-t}} \leq e^{t/2}\{\|u_{10} - u_{20}\|_{\Omega_R} +$$
$$+ \int_0^t e^{-s/2}\|f(u_1(s)) - f(u_2(s))\|_{\Omega_{R-s}} ds\}$$

---

[2]We essentially follow Ref.I. (quoted in footnote 4), to which we refer for a more detailed and general discussion.

$$\leq e^{t/2}\{\|u_{10} - u_{20}\|_{\Omega_R} +$$

$$+ \bar{C}(\Omega_R, \rho) \int_0^t e^{-s/2}\|u_1(s) - u_2(s)\|_{\Omega_{R-s}} ds\},$$

where $0 \leq t < R/2$ and

$$\rho = \sup_{0 \leq t < R/2} \|u_i(t)\|_{\Omega_{R-t}}, \quad (i = 1, 2).$$

Then, by Gronwall's lemma (see Appendix A, eq. (A.9))

$$\|u_1(t) - u_2(t)\|_{\Omega_{R-t}} \leq \exp\left[\left(\tfrac{1}{2} + \bar{C}(\Omega_R, \rho)\right) t\right] \|u_{10} - u_{20}\|_{\Omega_R}, \quad (\text{B.2})$$

which implies uniqueness and for $u_2 = 0$ yields the hyperbolic character.

4) We briefly sketch the idea of the proof. We first consider the case in which u_0 has compact support $\subset \Omega_R$, in which case the proof essentially reduces to a fixed point argument. Given $\rho > 0$, and a fixed u_0 with $\|u_0\|_{\Omega_R} < \rho/2$, we consider the operator S

$$(Su)(t) \equiv W(t)u_0 + \int_0^t W(t - s)f(u(s))ds \qquad (\text{B.3})$$

which maps $C^0(R, X_{loc})$ into itself (see 1) above). For T small enough, (depending on ρ), S is a contraction on the space

$$E(T, \rho) = \{u \in C^0([0, T], X_{loc}); \; \text{supp}\, u(t) \subset \Omega_{R+t};$$

$$\sup_{0 < t \leq T} \|u(t)\|_{\Omega_{R+t}} \leq \rho\},$$

which is complete with respect to the metric

$$d(u, v) = \sup_{0 \leq t \leq T} \|u(t) - v(t)\|_{\Omega_{R+t+1}}.$$

In fact, by using eq. (B.3), (u_0 fixed), the hyperbolic character of $W(t)$ and the local Lipschitz property of $f(u)$, one has, for $0 \leq t < T$, T small enough,

$$\|(Su)(t) - (Sv)(t)\|_{\Omega_{R+T+1}} \leq$$

$$\leq e^{t/2} \int_0^t ds\, e^{s/2} \bar{C}(\Omega_{R+T+1}, \rho) \|u(s) - v(s)\|_{\Omega_{R+T+1}} \leq$$

$$\leq e^{t/2}\, t\, \bar{C}(\Omega_{R+T+1}, \rho)\, d(u, v).$$

and S maps $E(T, \rho)$ into itself since

$$\|(Su)(t)\|_{\Omega_{R+T+1}} \leq e^{t/2} \left\{ \tfrac{1}{2}\rho + t\, \bar{C}(\Omega_{R+T+1}, \rho)\, \rho \right\} \leq \rho.$$

By Banach theorem on contractions, S has a fixed point which is the required solution in the interval $[0, T)$.

In the case in which u_0 does not have a compact support, we introduce a space cutoff putting

$$u_{0n} \equiv \begin{pmatrix} \chi_n \varphi_0 \\ \chi_n \psi_0 \end{pmatrix}, \quad \chi_n(x) \in C_0^\infty(\mathbb{R}^s),$$

$\chi_n(x) = 1$, if $|x| \leq n$, $\chi_n(x) = 0$ if $|x| \geq 2n$. Then, eq. (1.22) has a solution $u_n(t)$ by the previous argument.

Now, for any sphere Ω_{R-t}, by using the local Lipschitz condition and Gronwall's lemma, as in the derivation of eq. (B.2), we get

$$\|u_n(t) - u_m(t)\|_{\Omega_{R-t}} \leq \exp[(\tfrac{1}{2} + \bar{C}(\Omega_R, \rho)\, t\,]\|u_{0n} - u_{0m}\|_{\Omega_R}$$

and since u_{0n} converges in X_{loc} to u_0 as $n \to \infty$, also u_n converges in X_{loc} and it converges to the solution of eq. (4.6), with initial data u_0.

C   The global Cauchy problem

To prove theorem 4.1 we start by establishing the following a priori estimate.

Lemma C.1 *If the potential U is such that the local Lipschitz condition and the lower bound condition are satisfied, then any solution $u \in C^0([0,T], X_{loc})$ of eq. (4.6) with $supp_{0 \leq t < T} u(t) \subset \Omega_R$ satisfies*

$$\sup_{0 \leq t < T} \|u(t)\|_{\Omega_{R+1}} \equiv L < \infty. \tag{C.1}$$

Proof. The proof exploits the energy conservation

$$\frac{d}{dt}\{\tfrac{1}{2}\int_{\Omega_{R+1}} [(\nabla\varphi(t))^2 + (\psi(t))^2]d^s x + \int_{\Omega_{R+1}} U(\varphi(t))d^s x\} = 0. \tag{C.2}$$

(The above equation follows from the continuity equation for the energy momentum densities and the fact that there is no momentum flux through the boundary of Ω_{R+1}, since supp $u(t) \subset \Omega_R$.)

In fact, putting

$$K(t) \equiv \frac{1}{2}\int_{\Omega_{R+1}} [(\nabla\varphi(t))^2 + (\varphi(t))^2 + (\psi(t))^2]\, d^s x,$$

one gets from eq. (C.2)

$$K(t) = K(0) + \int_{\Omega_{R+1}} d^s x\, [U(\varphi(0)) - U(\varphi(t))] + \int_0^t dt' \int_{\Omega_{R+1}} \varphi(t')\psi(t')\, d^s x.$$

Now, by using the lower bound condition $(-U(\varphi(t)) \leq \alpha + \beta\varphi(t)^2)$ and the inequality $\varphi\ \psi \leq \tfrac{1}{2}(\varphi^2 + \psi^2) \leq (\varphi^2 + \psi^2 + (\nabla\varphi)^2)$, we have

$$K(t) \leq (K(0) + const) + (2\gamma + 1)\int_0^t dt' K(t').$$

Then, by Gronwall's lemma

$$K(t) \leq (K(0) + const)e^{(2\gamma+1)|t|}$$

which implies eq. (C.1).

Now, we can sketch the proof of Theorem 4.1. Any $u(\bar{t})$, $0 \leq \bar{t} < T$, defined by the solution for small times, for initial data of compact support, can be chosen as initial data for the equation

$$u(t) = W(t - \bar{t})\, u(\bar{t}) + \int_{\bar{t}}^{t} W(t - s)f(u(s))\, ds$$

equivalently for the equation

$$v(\tau) = W(\tau)\, v_0 + \int_0^{\tau} W(\tau - s)\, f(v(s))\, ds, \qquad \text{(C.3)}$$

where $v_0 \equiv u(\bar{t})$, $v(\tau) \equiv u(\tau + \bar{t})$, and by Lemma C.1 $\|v_0\|_{\Omega_{R+1}} < \rho$, $\rho > 2L$. Hence, the argument given in Appendix B can be applied and existence of solutions for eq. (C.3) can be proved for $0 \leq \tau < T_1$, with T_1 depending *only* on ρ. Since $\bar{t}$ can be chosen as close as we like to T this provides a continuation beyond T.

The existence of solutions for initial data with non-compact support is proved by the same argument, as at the end of Appendix B.

D The Cauchy problem for the non-linear wave equation with driving term

Theorem D.1 *The equation*

$$\delta(t) = W(t)\delta_0 + L(t) + \int_0^t W(t-s)\, g(\delta(s))\, ds, \qquad (D.1)$$

with $L(t)$, g, δ defined in Theorem 5.1, $L(0) = 0$, has a unique solution $\delta(t) \in C^0(\mathbb{R}, X)$, $X \equiv H^1(\mathbb{R}^s) \oplus L^2(\mathbb{R}^s)$.

Proof. Uniqueness follows from global Lipschitz continuity by the same argument of Appendix B, eq. (B.2), since the driving term $L(t)$ cancels. As in Appendix B, existence of solutions for small times follows by a fixed point argument applied to the space

$$E(T,\rho) = \{\delta \in C^0([0,T], X), \sup_{0 \le t \le T} \|\delta(t)\|_X < \rho\},$$

since

$$(S\delta)(t) \equiv W(t)\, \delta_0 + L(t) + \int_0^t W(t-s)\, g(\delta(s))\, ds$$

is a contraction on $E(T,\rho)$ for T small enough, $\|\delta_0\|_X < \rho/2$. Finally, the continuation beyond T is obtained as in Appendix C, by exploiting the a priori estimate

$$\sup_{0 \le t < T} \|\delta(t)\|_X \equiv L < \infty,$$

which follows from energy conservation $dE(t)/dt = 0$

$$E(t) = \tfrac{1}{2}\int d^s x [(\nabla\chi(t))^2 + (\zeta(t)+\psi_0)^2] - \int \chi(t)\, h\, d^s x + \int G(\chi(s))\, d^s x$$

(χ, ζ, h, G defined in Theorem 5.1). In fact, putting

$$H(t) \equiv E(t) + (\gamma + \tfrac{1}{2}) <\chi(t), \chi(t)> + \tfrac{1}{2} <\psi_0, \psi_0> + <\omega^{-1}h, \omega^{-1}h> \qquad (D.2)$$

where $< \cdot, \cdot >$ denotes the scalar product in L^2, $\omega = (-\Delta)^{\frac{1}{2}}$ and γ is the constant occurring in eq. (5.11), one has

$$H(t) = \tfrac{1}{4} < \omega\chi, \omega\chi > + \tfrac{1}{4} < \zeta, \zeta > + \tfrac{1}{2} < \chi, \chi > +$$

$$+ < \psi_0 + \tfrac{1}{2}\zeta, \psi_0 + \tfrac{1}{2}\zeta > + < \omega^{-1}h - \tfrac{1}{2}\omega\chi, \, \omega^{-1}h - \tfrac{1}{2}\omega\chi > +$$

$$+ \int [G(\chi(s)) + \gamma|\chi|^2]d^s x \geq$$

$$\geq \tfrac{1}{4}\|\delta\|_X \tag{D.3}$$

and

$$< \zeta + \psi_0, \zeta + \psi_0 > \leq 2\{< \psi_0 + \tfrac{1}{2}\zeta, \psi_0 + \tfrac{1}{2}\zeta > + \tfrac{1}{4} < \zeta, \zeta >\} \leq 2H \tag{D.4}$$

Hence,

$$H(t) = H(0) + 2(\gamma + \tfrac{1}{2}) \int_0^t d\tau < \chi(\tau), \zeta(\tau) + \psi_0 > \leq$$

$$\leq H(0) + 2(\gamma + \tfrac{1}{2})2 \int_0^t d\tau H(\tau),$$

so that, by eq. (D.3) and by Gronwall's lemma,

$$\frac{1}{4}\|\delta(t)\|_X \leq H(t) \leq H(0) \exp[4(\gamma + \frac{1}{2})t].$$

E Time independent solutions defining physical sectors

We briefly discuss the non-linear elliptic problem associated with the investigation of time independent solutions which define physical sectors (see ref. in footnote 23). For simplicity, we discuss the case $s \geq 3$. By the discussion of Sect. 6, we have to impose the condition $\nabla\varphi \in L^2(\mathbb{R}^s)$.

Proposition E.1 *Let us consider the non-linear elliptic problem* $(U \in C^2)$

$$\Delta\varphi - U'(\varphi) = 0, \quad \varphi \in H^1_{loc}(\mathbb{R}^s), \quad \nabla\varphi \in L^2(\mathbb{R}^s), \quad s \geq 3, \quad \text{(E.1)}$$

then,

i) the function $\tilde{\varphi}(r,\omega) \equiv \varphi(x)$, $x = r\omega$, $r > 0$, $\omega \in S^{s-1}$ *(the unit sphere of* $\mathbb{R}^s$*), is continuous in* r *and it has a finite limit* $\tilde{\varphi}(\infty,\omega)$ *as* $r \to \infty$*, for almost all* $\omega \in S^{s-1}$*, and the limit is independent of* ω*, briefly*

$$\lim_{|\vec{x}|\to\infty} \varphi(x) = \varphi_\infty, \quad \text{(E.2)}$$

ii) eq. (E.1), with boundary condition (E.2), does not have solutions unless φ_∞ *is a stationary point of the potential*

$$U'(\varphi_\infty) = 0, \quad \text{(E.3)}$$

iii) if φ_∞ *is an absolute minimum of* U*, then* φ *is the unique solution of (E.1) with* φ_∞ *as boundary value at infinity and* $\varphi = \varphi_\infty$*.*

Proof i) By using a mollifier technique, one reduces the proof of the existence of the limit $\tilde{\varphi}(\infty,\omega)$, for almost all $\omega \in S^{s-1}$, to the estimate

$$|\tilde{\varphi}(r,\omega) - \tilde{\varphi}(r_0,\omega)| \leq \int_{r_0}^r \left|\frac{d}{dr'}\tilde{\varphi}(r',\omega)\right| dr' \leq$$

$$\leq \left(\int_{r_0}^r \left| \frac{d\varphi}{dr'} \right|^2 (r')^{s-1} dr' \right)^{\frac{1}{2}} \left(\int_{r_0}^r (r')^{s-1} dr' \right)^{\frac{1}{2}} \leq$$

$$\leq const \left(\int_{r_0}^r \left| \vec{\nabla} \varphi \, \frac{\vec{x}}{r'} \right|^2 (r')^{s-1} dr' \right)^{\frac{1}{2}} |r^{2-s} - r_0^{2-s}|^{\frac{1}{2}}.$$

$$(E.4)$$

The independence of ω, for almost all ω, follows from the following fact: if φ is locally measurable and $\nabla\varphi \in L^p(\mathbb{R}^s)$, $1 \leq p \leq s$, then there exists a constant A, depending on f, such that

$$\varphi - A \in L^q(\mathbb{R}^s), \qquad \frac{1}{q} = \frac{1}{p} - \frac{1}{s}. \qquad (E.5)$$

To see this, we define

$$\mathcal{H}^L = \{ f \in \mathcal{S}'(\mathbb{R}^s), \nabla f \in L^p(\mathbb{R}^s) \}$$

and associate to each element of $\mathcal{H}^L$ the norm

$$\|f\|_{\mathcal{H}^L} = \|\nabla f\|_{L^p}.$$

The so obtained normed space is complete, i.e. if $f_j \in \mathcal{H}^L$ is a Cauchy sequence, then $\nabla_k F_j$ converges to an $F^{(k)} \in L^p$ and since $\nabla_k F^j - \nabla_j F^{(k)} = 0$ in the sense of distributions, there exists an f such that $F^{(k)} = \nabla_k f$. It is convenient to consider the quotient space $\mathcal{H} = \mathcal{H}^L / \mathcal{H}_0$, where $\mathcal{H}_0 = \{ f \in \mathcal{S}'(\mathbb{R}^s), \nabla f = 0 \}$. $C_0^\infty(\mathbb{R}^s)$ is weakly dense in $\mathcal{H}$, i.e. if $h \in (\mathcal{H})^*$, the dual space of $\mathcal{H}$, then $h(g) = 0$, $\forall g \in C_0^\infty(\mathbb{R}^s)$, implies $h = 0$; in fact if h is a continuous linear functional on $\mathcal{H}$

$$|h(g)| \leq const \|\nabla g\|_{L^p}$$

and by the Riesz representation theorem this implies that there exists a $h \in L^q$ such that

$$h(g) = \int h \nabla g \, d^s x.$$

Hence, $h(g) = 0$, $\forall g \in C_0^\infty(\mathbb{R})$, implies $0 = \int h\nabla g d^s x = -\int \nabla h g d^s x$, i.e. $\nabla h = 0$, i.e. $h = const$, i.e. $h = 0$ as a functional on $\mathcal{H}$.

Finally if $f \in \mathcal{H}$, there exists a sequence $\{f_j \in C_0^\infty(\mathbb{R}^s)\}$ with $f_j \to f$ in $\mathcal{H}$; this implies that ∇f_j converges in $L^p(\mathbb{R}^s)$ and, by Sobolev inequality

$$\|f_j\|_{L^q} \leq const\|\nabla f_j\|_{L^p},$$

$f_j \to \tilde{f}$ in L^q and $\tilde{f}$ belongs to the same equivalence class of f, i.e. $f = \tilde{f} + const$.

ii) Since $\tilde{\varphi}(r,\omega)$ is continuous in r and $U \in C^2$

$$\lim_{r\to\infty} U'(\tilde{\varphi}(r,\omega)) = U'(\tilde{\varphi}(\infty,\omega)) = U'(\varphi_\infty).$$

Eq. (E.1) implies that also $\lim_{r\to\infty} \Delta\tilde{\varphi}(r,\omega)$ exists and it is independent of ω. Furthermore, $\forall f(r) \in \mathcal{D}(\mathbb{R}^+)$, with $\int_0^\infty f(r)dr = 1$

$$U'(\varphi_\infty) = \lim_{r\to\infty} \Delta\tilde{\varphi}(r,\omega) = \lim_{a\to\infty}(\Delta\tilde{\varphi})(r+a,\omega) =$$
$$= \lim_{a\to\infty} \int_0^\infty dr f(r)(\Delta\tilde{\varphi})(r+a,\omega) =$$
$$= \lim_{a\to\infty} \int_0^\infty dr(\Delta f(r))\tilde{\varphi}(r+a,\omega) =$$
$$= \varphi(\infty) \int_0^\infty dr \Delta f(r) = 0.$$

iii) If φ_∞ is an absolute minimum

$$\tilde{U}(\varphi) \equiv U(\varphi) - U(\varphi_\infty) \geq 0$$

and the solutions of eq. (E.1) are stationary points of the functional

$$H(\varphi) = \int [|\nabla\varphi|^2 + \tilde{U}(\varphi)]d^s x.$$

Now, by putting $\varphi_\lambda(x) \equiv \varphi(\lambda x)$, $\lambda \geq 0$, we get

$$H_\lambda = \int [|\nabla\varphi_\lambda|^2 + \tilde{U}(\varphi_\lambda)]d^s x = \int [\lambda^{-1}|\nabla\varphi|^2 + \lambda^{-3}\tilde{U}(\varphi)]d^s x$$

and

$$\delta^{(\lambda)} H = \delta\lambda\frac{\partial H_\lambda}{\partial\lambda} = -(\delta\lambda)\lambda^{-2}\int[|\nabla\varphi|^2 + 3\lambda^{-2}\tilde{U}(\varphi)]d^s x.$$

Hence, the condition of stationarity and the positivity of $\tilde{U}$ yield

$$\int|\nabla\varphi|^2 d^s x = 0, \quad \text{i.e. } \nabla\varphi = 0, \quad \text{i.e. } \varphi = \varphi_\infty,$$

and $\tilde{U}(\varphi_\infty) = 0$.

Elenco dei volumi della collana
"Appunti"
pubblicati dall'Anno Accademico 1994/95

GIUSEPPE BERTIN (a cura di), *Seminario di Astrofisica,* 1995.

EDOARDO VESENTINI, *Introduction to continuous semigroups,* 1996.

LUIGI AMBROSIO, *Corso introduttivo alla Teoria Geometrica della Misura ed alle Superfici Minime,* 1997.

CARLO PETRONIO, *A Theorem of Eliashberg and Thurston on Foliations and Contact Structures,* 1997.

MARIO TOSI, *Introduction to Statistical Mechanics and Thermodynamics,* 1997.

MARIO TOSI, *Introduction to the Theory of Many-Body Systems,* 1997.

PAOLO ALUFFI (a cura di), *Quantum cohomology at the Mittag-Leffler Institute,* 1997.

GILBERTO BINI, CORRADO DE CONCINI, MARZIA POLITO, CLAUDIO PROCESI, *On the Work of Givental Relative to Mirror Symmetry,* 1998

GIUSEPPE DA PRATO, *Introduction to differential stochastic equations,* 1998

HERBERT CLEMENS, *Introduction to Hodge Theory,* 1998

HUYÊN PHAM, *Imperfections de Marchés et Méthodes d'Evaluation et Couverture d'Options,* 1998

MARCO MANETTI, *Corso introduttivo alla Geometria Algebrica,* 1998

AA.VV., *Seminari di Geometria Algebrica 1998-1999,* 1999

ALESSANDRA LUNARDI, *Interpolation Theory,* 1999

RENATA SCOGNAMILLO, *Rappresentazioni dei gruppi finiti e loro caratteri,* 1999

SERGIO RODRIGUEZ, *Symmetry in Physics,* 1999

F. STROCCHI, *Symmetry Breaking in Classical Systems* and Nonlinear Functional Analysis, 1999

"CompoMat" Loc. Braccone, 02040 Configni (RI), Italy
Finito di stampare 6 nel maggio 2000